Wire Arc Additive Manufacturing

This book presents wire arc additive manufacturing (WAAM), its variants, processing steps, and the mechanical and microstructural aspects of developed components, along with a logical sequence that provides a ready reference.

Wire Arc Additive Manufacturing: Fundamental Sciences and Advances introduces the timeline and history of WAAM. It offers a critical review of WAAM, its variants, processing steps, and the mechanical and microstructural aspects of developed components. The book showcases the methods and practices that need to be followed to synchronise WAAM with other conventional metal additive manufacturing, as well as other conventional techniques. The process steps, equipment, and the different materials used are discussed in detail, along with the various process parameters and their optimisation to counter the challenges and defects. Applications, trends, and case studies are also included in the book.

This book, aimed at providing a concrete reference for researchers, academics, and professionals, allows for a thorough understanding of the different concepts and intricacies of WAAM.

Advanced Manufacturing of Materials: Fundamentals, Advancements, and Applications
Sandeep Rathee, Manu Srivastava, Manoj Gupta

Wire Arc Additive Manufacturing: Fundamental Sciences and Advance
Sandeep Rathee, Manu Srivastava

Wire Arc Additive Manufacturing

Fundamental Sciences and Advances

Edited by
Sandeep Rathee
Manu Srivastava

CRC Press
Taylor & Francis Group
Boca Raton London New York

CRC Press is an imprint of the
Taylor & Francis Group, an **informa** business

First edition published 2024
by CRC Press
6000 Broken Sound Parkway NW, Suite 300, Boca Raton, FL 33487-2742

and by CRC Press
4 Park Square, Milton Park, Abingdon, Oxon, OX14 4RN

CRC Press is an imprint of Taylor & Francis Group, LLC

ISBN: 978-1-032-41914-5 (hbk)
ISBN: 978-1-032-42599-3 (pbk)
ISBN: 978-1-003-36341-5 (ebk)

DOI: 10.1201/9781003363415

Typeset in Times
by SPi Technologies India Pvt Ltd (Straive)

Contents

Preface

May the knowledge rest in the hands of the one most worthy.

Sandeep Rathee and Manu Srivastava

Exchanging ideas and effectively communicating the accumulated knowledge is a moral responsibility of every dedicated researcher and academician. Life is unpredictable, and while we can, we should make every effort to share our learnings because this is the only asset that cannot be transferred to the one we want and does not follow the doctrine of inheritance. Our future generation of researchers should benefit from our learnings and be saved from the challenges our generation has faced to build upon something more worthwhile without the need to go into the same iterative cycles wherever possible with a vision to make a technologically and socially strong community based on deep-rooted foundations. With this thought process in place, the editors have come together to disseminate information gathered from years of experience in the field of wire arc additive manufacturing (WAAM) with interested and willing readers. Today, a wide variety of literature in the form of journal articles is available in the field of WAAM, but most of these are more or less confined only to some particular areas. A resource that presents the overall picture in the area of WAAM is very much required. This book is a novel venture toward the said direction. It is ensured to present details in simple yet precise language with clarity to cater to a wide variety of readers globally.

Metal additive manufacturing (MAM) is a term coined for utilising metals and their alloys as the raw material for additive manufacturing. Directed energy deposition (DED) is one special class of MAM techniques. MAM techniques have basically paved the way for hybrid additive manufacturing (HAM) methods that utilise the amalgamation of conventional manufacturing methods and additive manufacturing. WAAM is a special HAM technique that utilises the basic fundamentals of the DED technique along with conventional welding and CNC milling techniques for fabricating parts. It is one of the fastest-growing processes that enables the repair of parts, as well as the development of large-sized parts with acceptable levels of accuracy. Its key features include several outstanding superiorities, such as an elevated rate of depositing materials, relatively low equipment investment costs, and different feedstock materials available in the market. It fabricates large components with a high deposition rate and less material wastage; the operating and setup cost of WAAM technology is lower than laser and electron beam technology.

The chapters included in this book have been briefly introduced here to make the reader well versed in the overall content.

Chapter 1 presents an overview of additive manufacturing (AM) and introduces the readers to the vast variety of applications for AM in industrial areas such as aircraft, biomedicine, automobiles, and so on. The goal of this chapter is to educate the reader about AM technology, including its history, key operations, processing techniques, and applications.

Chapter 2 presents a comprehensive overview of the WAAM process and a case study on the metallurgical characterisation of as-WAAMed mild steel parts. The chapter provides a fundamental understanding of WAAM in the manufacture of metallic parts and makes recommendations when applying this technology in practical applications.

Chapter 3 introduces the process planning criterion in WAAM, starting with an overview of process planning parameters and their selection strategies. End-use components need to be machined using appropriate machining methods. Several process selection parameters and computer-aided process planning need to be addressed to establish the capability of WAAM to produce near-perfect components of particular materials. Primary process selection criteria include CAD model parameters, tool path generation, welding process, welding process parameters, shielding gas, and wire selection.

Chapter 4 introduces and discusses details of metal inert gas wire arc additive manufacturing (MIG-WAAM), which is a type of DED technology that is used to create complex and intricate metal components by melting and depositing metal in a layered fashion. It can produce large, complex parts quickly and efficiently, with minimal waste. It is also versatile and can work with a wide range of metals, including steel, titanium, aluminium, and nickel alloys. MIG-WAAM has applications in various industries, including aerospace, automotive, marine, etc. It can be used to create parts such as engine components, turbine blades, and orthopaedic implants.

Chapter 5 presents details of the cold metal transfer-based additive manufacturing technique, which is a new promising approach based on wire-feed AM. It is gaining more popularity than its contemporary AM processes for MAM due to its capability of economically producing large-sized components with relatively high deposition rates and lower heat input. This chapter introduces cold metal transfer-based wire arc additive manufacturing (CMT-WAAM), starting with an overview of CMT-WAAM and the detailed mechanism of the CMT process, and its selection preference over the other variants of WAAM. A critical review of the microstructure and mechanical properties of various metals and alloys fabricated through the CMT-WAAM technique has been reported. The advantages of CMT-WAAM have piqued the interest of many industrial experts and researchers for further developments in this technique; thus, the recent advances performed in this sector have been summarised in the last section of this book.

Chapter 6 provides the knowledge necessary to accommodate the current and future material for WAAM. It is a common perception that during WAAM, the wire is fed through a welding torch, where it is melted and deposited onto the workpiece using an electric arc. The paradigm of wired-type material, however, is beginning to change in favour of powder or a hybrid of the two types of material as a result of some new research. In relation to the past, present, and future trends, this chapter specifically addresses the different materials used for the WAAM process. Additionally, the chapter covers various facets of the raw material form (wires, powders, or hybrids) in accordance with the optimisation of mechanical and functional properties and their respective metallographic representations.

Chapter 7 attempts a detailed study of AM, WAAM, and current challenges in WAAM of non-ferrous alloys, particularly aluminium, magnesium, and titanium alloys.

Chapter 8 presents an overview of the different aspects related to the fabrication of functionally graded materials (FGMs) via MAM techniques, including WAAM. Different processing techniques and their challenges are discussed in detail.

Chapter 9 explores the impact of manufacturing parameters of the WAAM process on the surface quality of parts produced via this AM method through the lens of existing research works. The effect of parameters associated with the CAD design, tool path configuration, welding process and instrument settings, shielding gas, and wire materials were examined, and the part they play in the development of defects like porosity, surface waviness, residual stress, deformation, crack, delamination, and geometric accuracy. In conclusion, the methods that have been adopted for the mitigation of these defects are highlighted in detail in this chapter.

Chapter 10 documents the significant influence of hot isostatic pressing, heat treatment techniques, machining process, and surface modification techniques as they enhance the structural integrity performance of WAAM-processed Ti-6Al-4V components. This chapter systematically reviews the recent progress in post-processing techniques as they improve the material properties of Ti-6Al-4V components produced by the WAAM process. WAAM components fabricated with the use of titanium and its alloys have extensively gained awareness in biomedical, marine, and aerospace industries owing to their mechanical properties, biocompatibility, and excellent corrosion resistance properties. Post-processing is an essential technique used for fine-tuning microstructure and material properties, correcting incurred errors, and improving surface properties of WAAM-processed Ti-6Al-4V alloy to enhance the quality of metallic parts.

Chapter 11 presents a case study examining the mechanical properties of WAAM-fabricated, multi-layer wall structures. First, wrought mild steel was used to perform tensile and hardness tests then tests were performed on the top, middle, and bottom zones of the WAAM-fabricated, multi-layered wall. Results showed that tensile strength at the top zone has higher values than the middle and bottom zone. Similarly, hardness at the bottom zone has the highest value compared to other zones. The findings of this study showed that the mechanical properties of multi-layer wall structures constructed using WAAM might vary based on the precise zone of the component under consideration. However, the findings showed WAAM's promise as a productive and economical manufacturing technique for creating substantially intricate components with high strength and hardness.

The quantum of information related to different WAAM techniques, fast degree of obsolescence, and extremely high levels of ongoing technical as well as technological advances puts a restraint on presenting details of every aspect of each related topic. The editor group has, however, put in their best efforts to make this book informative and interesting. This book is a result of dedicated research in the field of WAAM, as well as collaboration with different peer groups and an in-depth literature review. Editors most sincerely hope that the book is a valued knowledge source for upcoming research groups, academia, and industry. It is advised to apply the information in this book to promote research and development in the field of WAAM.

All queries, advice, and observations regarding the book are most welcome.

Dr. Sandeep Rathee
Dr. Manu Srivastava

Acknowledgements

At the start, our editor, Mrs. Cindy Renee Carelli, for being such a huge pillar of strength, needs wholehearted acknowledgement. The editors profusely thank the entire team of CRC Press for their support towards the initiative of bringing forth this book.

The editors thank all the contributing authors of different chapters for sharing their expertise and helping us in bringing the book to its present form. The editors wholeheartedly thank their respective institutions, PDPM Indian Institute of Information Technology, Design, and Manufacturing in Jabalpur and National Institute of Technology Srinagar, Jammu and Kashmir. We express deep gratitude to them for their valuable support toward the endeavour to come up with the present edited book.

Dr. Sandeep Rathee wishes to thank his mentors, Dr. Pulak M. Pandey and Dr. Sachin Maheshwari. He also acknowledges the support of his students and family, especially his parents, Shri Raj Singh Rathee and Smt. Krishna Rathee, for their constant support.

Dr. Manu Srivastava wishes to thank her mentors. She acknowledges the support of her students for their inspiration and support. Projects like this need a lot of dedicated effort, which often comes at the cost of time kept apart from the family. No words of gratitude would suffice to acknowledge her family for their affectionate support and unfaltering trust.

Authors devote and dedicate this work to the Divine Creator based upon their belief that the strength to bring any thoughtful and noble endeavour into being emerges from the Almighty with a wish that this work makes a valuable addition for its readers.

Dr. Sandeep Rathee
Dr. Manu Srivastava

Editors

Dr. Sandeep Rathee is currently serving the Department of Mechanical Engineering, National Institute of Technology Srinagar, India, as an assistant professor. His previous assignment was as a Post-Doctoral Fellow at the Indian Institute of Technology Delhi (IIT Delhi). He is the recipient of the prestigious National Post-Doctoral Fellowship from the Science and Engineering Research Board (SERB; Government of India). Prior to this, he worked with the Amity School of Engineering and Technology, Amity University Madhya Pradesh, India. He was awarded a Ph.D. degree from the Faculty of Technology, University of Delhi. His field of research mainly includes friction stir welding/processing, advanced materials, composites, AM, advanced manufacturing processes, and characterisation. He has authored over 60 publications in various international journals of repute and refereed international conferences. He has been awarded eight industrial design patents and one copyright. He has authored/edited seven books in the field of advanced manufacturing. He is working as an editor-in-chief of a book series titled "Advanced Manufacturing of Materials" with CRC Press, Taylor & Francis Group. He has also worked as managing guest editor for a special issue of a Scopus-indexed Elsevier journal. He is associated with several reputed journals in the capacity of editorial member. Additionally, he is serving as a reviewer for more than 30 journals. He has completed/is handling externally funded projects of more than 07 Million INR in the field of AM. He has supervised/is supervising 4 Ph.D. scholars, 2 M.Tech, and 20 B.Tech students. His works have been cited more than 1,250 times (as per Google Scholar).

He has a total teaching and research experience of more than ten years. He has delivered invited lectures and chaired scientific sessions at several national and international conferences, STTPs, and QIP programmes. He is a life member of the Additive Manufacturing Society of India (AMSI) and VIBHA.

Dr. Manu Srivastava is presently serving PDPM Indian Institute of Information Technology, Design, and Manufacturing in Jabalpur, India, in the Department of Mechanical Engineering. Her previous assignment was as a professor and head of the Department of Mechanical Engineering and director of Research, Faculty of Engineering and Technology, MRIIRS, Faridabad, India. She has completed her Ph.D. in the field of additive manufacturing (AM) from the Faculty of Technology, University of Delhi. Her field of research is AM, friction-based AM, friction stir processing, advanced materials, manufacturing practices, and optimisation techniques. She has authored/co-authored around 100 publications in various technical platforms of repute. She has been awarded 18 industrial design patents and 1 copyright.

She has authored/edited seven books in the field of advanced manufacturing. Out of these, three have already been published with CRC Press, Taylor & Francis Group, and others are either in press or in an accepted stage. Two of them are with Wiley, one with Elsevier and one with Springer Publishing Company.

She is working as an editor-in-chief of a book series titled "Advanced Manufacturing of Materials" with CRC Press, Taylor & Francis Group. She has also worked as a guest editor for a special issue of a Scopus-indexed Elsevier journal. She is on the editorial board of several reputed journals in the capacity of editorial member. Additionally, she is serving as a reviewer for more than 30 journals. She is working on various projects of around 5 million INR and has completed several consultancy works funded by the Government of India in the field of hybrid additive manufacturing and rehabilitation robotics. She has a total teaching and research experience of around 15 years. She has won several proficiency awards during the course of her career, including merit awards, best teacher awards, etc. One special award that needs to be mentioned is the Young Leader Award. She has delivered invited lectures and chaired scientific sessions at several national and international conferences, short-term training programme (STTPs), and quality improvement programme (QIP) programmes. She is a life member of the Additive Manufacturing Society of India (AMSI), Vignana Bharti (VIBHA), the Institution of Engineers (IEI India), Indian Society for Technical Education (ISTE), Indian Society of Theoretical and Applied Mechanics (ISTAM), and Indian Institute of Forging (IIF).

Contributors

Sheikh Nazir Ahmad
Department of Mechanical Engineering
National Institute of Technology Srinagar
Jammu & Kashmir, India

Tat Khoa Doan
Advanced Technology Center
Le Quy Don Technical University
Hanoi, Vietnam

Peter Kayode Farayibi
Department of Industrial and Production Engineering
Federal University of Technology
Akure, Nigeria

Prashant K. Jain
FFF Lab, Mechanical Engineering Discipline
PDPM Indian Institute of Information Technology, Design, and Manufacturing
Jabalpur, India

Basant Kumar
Department of Mechanical Engineering
National Institute of Technology Srinagar
Jammu & Kashmir, India

Himanshu Kumar
Laboratory for Advanced Manufacturing and Processing
Indian Institute of Technology Jammu
Jammu, India

Van Thao Le
Advanced Technology Center
Le Quy Don Technical University
Hanoi, Vietnam

Manidipto Mukherjee
CSIR-Central Mechanical Engineering Research Institute
Durgapur, India
and
Academy of Scientific and Innovative Research
Ghaziabad, India

Muhammad Mursaleen
Department of Mechanical Engineering
National Institute of Technology Srinagar
Kashmir, India

Shazman Nabi
Department of Mechanical Engineering
National Institute of Technology Srinagar
Kashmir, India

Ikeoluwa Ireoluwa Ogedengbe
Department of Industrial and Production Engineering
Federal University of Technology
Akure, Nigeria

Temitope Olumide Olugbade
Department of Industrial and Production Engineering
Federal University of Technology
Akure, Nigeria

Babatunde Olamide Omiyale
Department of Industrial and Production Engineering
Federal University of Technology
Akure, Nigeria

Amos Babatunde Osasona
Department of Industrial and Production Engineering
Federal University of Technology
Akure, Nigeria

Amrit Raj Paul
Royal Melbourne Institute of Technology
Melbourne, VIC, Australia
and
CSIR-Central Mechanical Engineering Research Institute
Durgapur, India
and
Academy of Scientific and Innovative Research
Ghaziabad, India

Sandeep Rathee
Department of Mechanical Engineering
National Institute of Technology Srinagar
Jammu & Kashmir, India

S. Shiva
Laboratory for Advanced Manufacturing and Processing
Indian Institute of Technology Jammu
Jammu, India

Amritbir Singh
Laboratory for Advanced Manufacturing and Processing
Indian Institute of Technology Jammu
Jammu, India

Manu Srivastava
Hybrid Additive Manufacturing Lab
Department of Mechanical Engineering
PDPM Indian Institute of Information Technology, Design, and Manufacturing
Jabalpur, India

Bunty Tomar
Laboratory for Advanced Manufacturing and Processing
Indian Institute of Technology Jammu
Jammu, India

M. F. Wani
Department of Mechanical Engineering
National Institute of Technology Srinagar
Kashmir, India

Ashish Yadav
Hybrid Additive Manufacturing Lab
Department of Mechanical Engineering
PDPM Indian Institute of Information Technology, Design, and Manufacturing
Jabalpur, India

Juneed Yawar
Department of Mechanical Engineering
National Institute of Technology Srinagar
Kashmir, India

1 Additive Manufacturing *Introduction, Fundamentals, Types, and Applications*

Amritbir Singh, Himanshu Kumar, and S. Shiva
Indian Institute of Technology Jammu, Jammu, India

1.1 INTRODUCTION

To be competitive in today's global economy, manufacturers must adapt to new technologies and improve their production processes. As production shifts from mass manufacturing to mass customisation, improvements to equipment and processes are needed to accommodate more product variety, shorter life spans of products, and lower batch sizes. Therefore, the concept of additive manufacturing (AM) was put forward to deal with the change in the demand of customers over time. Owing to this, the AM processes are at the vanguard of Industry 4.0, which is the focus of numerous sectors. One of the factors that have contributed to this technique being more successful and playing a larger part in the production sector is the emergence of the digitalisation of manufacturing industries.

The potential of AM to rapidly generate items with complicated forms and detailed features without expensive tooling or lengthy lead times has transformed the manufacturing industry [1]. More and more manufacturing facilities are realising the benefits of using AM technology. By eliminating unnecessary steps and decreasing the need for human labour, AM speeds up production and improves accuracy. The absence of necessary tools also makes it simpler to tailor items to individual users or uses, increasing satisfaction. The notion of mass production has guided factory layouts for decades. The logic behind this strategy is that mass-producing one good is more cost-effective than mass-producing many different goods. AM enables the fabrication of highly personalised items in small quantities, as opposed to the mass production that is possible with conventional manufacturing methods. Therefore, factories might be built to be more adaptable and rapid to meet shifting customer demands. Furthermore, AM might allow for more efficient manufacturing processes, with reduced waste and smaller errors.

Taking into view the benefits of AM, the concerned sector has experienced tremendous growth in the technological sphere during the past 40 years. Earlier, AM was only utilised for a limited number of scientific applications (such as prototyping), but currently, it has grown with regard to feedstock and applications. Chuck Hull, a co-founder of 3D Systems, originally obtained a patent for the technique in 1984 and then began commercialising it in 1987 [2]. After the first industrial manufacturing technology was introduced four years ago, fused deposition modelling

DOI: 10.1201/9781003363415-1

(FDM) [3] and STEREOS [4] helped the additive technology industry expand. However, the machine's limited material variety has spurred a number of scholars to concentrate on increasing their material variability. As a consequence, EOS introduced the first AM system for metal processing (EOSINT M160). A powder mixture was used as the base material, and low melting point components served as the adhesive that bound particles with high melting temperatures. Figure 1.1 depicts the historical chronology of several businesses investing in this technology. Moreover, the research on the AM market projected that 43 years after the introduction of the first commercialised technology, the industry had broadened to €13.4 billion, expanding at a pace of 22% annually. More than 200 companies are actively striving to create cutting-edge technology, software, and materials, indicating the industry's continued viability. The functionality of AM systems is dramatically improving as a result of the rapid rate of development. Newer machines are able to produce overhanging components without a requirement for complex support structures, and they can make stronger components by managing the position of fibre reinforcements with magnetic forces. High-strength metal alloys and medical-grade polymers are two examples of the growing variety of materials compatible with AM systems. Therefore, it becomes paramount for the budding AM enthusiast to take a dip into this

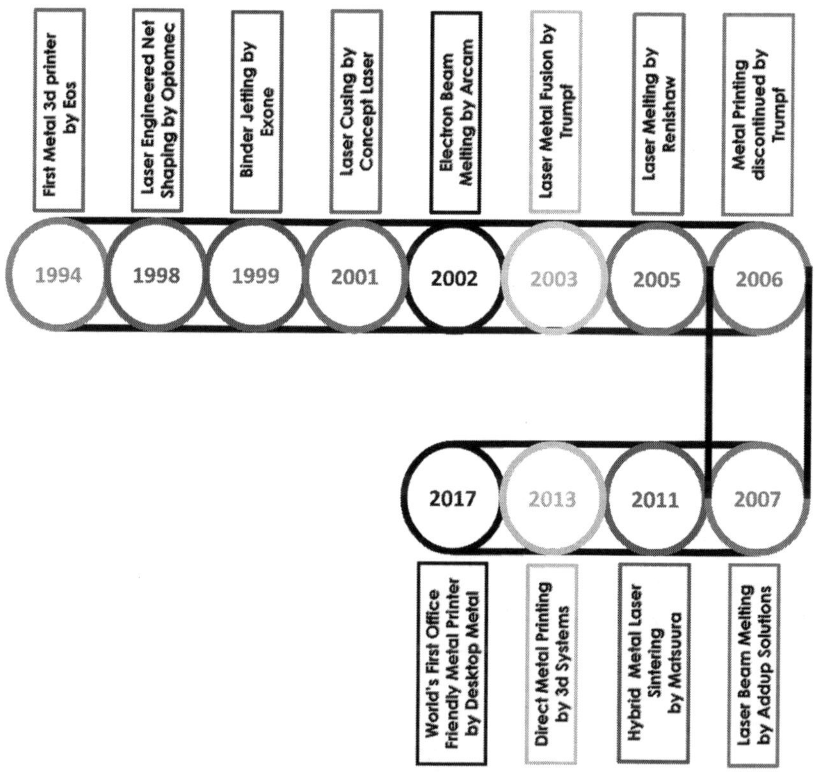

FIGURE 1.1 The chronological sequence of the various companies' entrance into the 3D printer industry under their individual exclusive brand names of the AM method.

technology by understanding the basics related to the same. Hence, in this chapter, the authors will discuss many aspects of AM, from the procedural steps involved during AM processing, technology's benefits and drawbacks, and the applications that are making use of the process.

1.2 STEP-BY-STEP PROCEDURE OF AM

Over the last several years, the highest level of the production chain in metal AM has evolved from generating prototypes to manufacturing finished parts. Furthermore, both the flexibility of the AM process and the variety of materials it may use have seen tremendous improvements. Product development, on the other hand, is often carried out by executing a sequence of key processes to achieve the build-up of stated materials. Regardless of the equipment used, these broad stages are required to generate the final output and are constant with all 3D printing procedures [5]. The procedure would be divided into six major steps and described as follows.

1.2.1 CAD File Preparation

A computer-aided design (CAD) model for an AM process is needed in each situation, regardless of whether the object is a prototype or a finished product. Additionally, the development of AM techniques wouldn't have proven feasible without 3D modelling. Only after learning how to examine solid structures using software in a computing device were researchers able to develop technology to physically duplicate them. Visualising the appearance of the product is the first step in the AM technique of product manufacturing. The design of the component will initially be represented by hazy sketches with vague dimensions. Once the concept has been translated into proper drawings and dimensions, it is deemed ready for translation to digital form using various software [6]. SOLIDWORKS, CATIA, FUSION 3600, CREO, AUTOCAD, and other CAD software products are extensively used within the wider banner of CAD programmes. An open-source tool, on the contrary, such as TINKERCAD, is accessible as a web-based application and is regarded as simple to practice for beginners [7]. A 3D scan of an actual component is a substitute approach for creating digital data. The most successful techniques to aid the creator in the replication operation are 3D scanning and photogrammetry tools. Using CAD tools or 3D scanning to create digital representations is feasible by storing the information in a particular format. The first company to publicise this standard file in AM technology was 3D Systems, which called it Standard triangular Language (STL) [1]. STL is the name given to this language because it is most frequently used to represent a surface area in the form of small triangles.

1.2.2 Pre-processing of Design

Prior to printing, a number of steps must be taken once the STL file has been created. This preliminary processing falls under the category of critical significance known as Design for Additive Manufacturing (DfAM). As a result, different corporations provide distinct software solutions for making use of this particular concept of

DfAM. The component that is to be fabricated is digitally placed in the build volume during this stage of AM. Support structures become unavoidable when a component is oriented at a specific angle. These structures are used in almost all metal AM techniques. Their inclusion serves to prevent component deformation caused by residual stress, improve the printing capability of overhang characteristics, and guarantee proper heat conduction. However, their utilisation is frequently seen as troublesome in terms of the aesthetics of the component. Furthermore, in terms of software adaptability, the designer may realistically put the number of components to be produced simultaneously within the construction volume. Additionally, these software are present to ensure that the part has been printed as effectively as feasible. The term "efficiency" in this context refers to enhanced qualities, cost, and time savings achieved by optimising process parameters [8].

1.2.3 Slicing of the Part

Following the implementation of pre-processing methods, the component is cut into the desired amount of 2D sections using the software. To do this, slicing divides the object into numerous levels. In essence, it gives adequate detail regarding the path to be taken by the heat source for each layer. Such data is generated as G-codes and is therefore comprehensible by the AM equipment. Slicing a 3D model, in other words, implies that the blueprint can be understood by the AM machine [1].

1.2.4 Machine Configuration

Every AM equipment has set-up settings that are unique to that system. These characteristics, particularly in AM, define the level of quality of the part created for the final use. As a result, optimum variables selection of particular materials is dominating in part superiority. In rare cases, despite an incorrect set-up parameter, a part can be constructed. However, the end result in terms of component quality may be unacceptably low. There are additional system starting stages that need to be carried out after putting the STL file into the AM technology. The majority of these activities involve setting up the AM machine to build the part. The maker must ensure that the equipment receives sufficient raw material to complete the building operation. The machine's set-up has to account for more than simply material feeding; the oxygen content in the chamber has to be maintained within safe parameters. Because of the tendency of the raw material or molten pool to undergo oxidation, an inert gas atmosphere becomes essential. To reduce oxidation effects, helium, nitrogen, argon, or their mixes can be used. As a result, the amount of pressure in the gas tanks must be monitored before the procedure begins [9].

1.2.5 Build-Up Process

The machinery used to make the component is basically an automated system that is capable of doing the critical task. As a result, it may operate without supervision for the vast majority of the time. However, a check of the material amount power supply is still required from time to time to ensure appropriate operating procedures.

1.2.6 Component Separation and Post-Processing

The component needs to be taken out of the base plate after the formation procedure is finished. The component is believed to be held within a layer of particles (especially for powder bed fusion (PBF)), but before it is extracted, it is cleaned with a brush or a vacuum cleaner. Following the removal of superfluous powder, the support is removed using hand tools or machines. As a result, the component face linked to the support has poor surface quality [10]. This effect makes post-processing obvious. Such processing entails refining, sandpapering, sealing, and other finishing procedures [11]. Furthermore, AM components are manufactured to fulfil the implementation requirements. As a result, the part is frequently considered unusable in its as-formed (straight off the machine) condition and requires extra processing, like heat treatment.

1.3 BENEFITS OF AM

AM is not a novel concept. Numerous more sectors have embraced the advantages of AM in the past few years as the process behind it has advanced and prices have decreased (Figure 1.2). As AM has become more prevalent, technology may be utilised in a number of ways. So, let us examine why AM has proven so influential.

1.3.1 Reduced Start-Up Costs

Manufacturing start-up costs might be too expensive. The requirement to design unique tooling for each novel product you desire to make might restrict the range of what is financially possible to produce. Commercial AM equipment costs around 1,000 pounds, whereas home or hobbyist solutions cost a few hundred pounds.

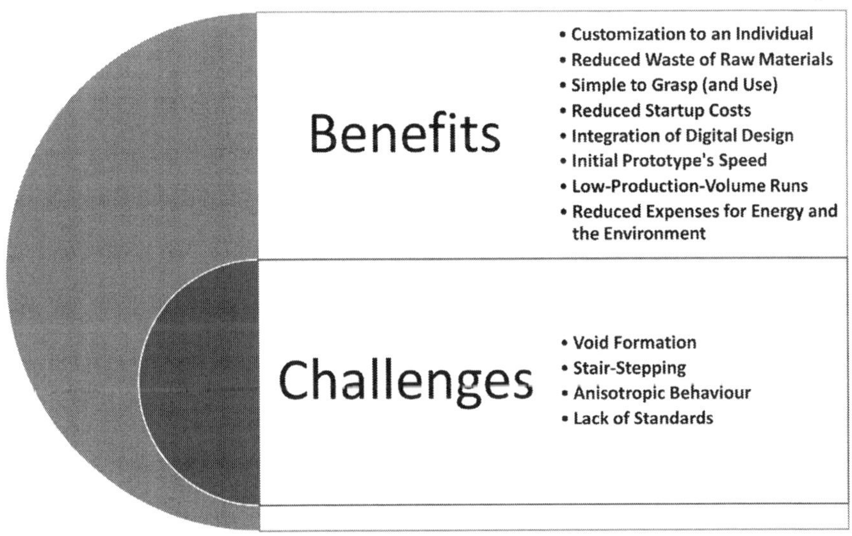

FIGURE 1.2 The benefits and challenges of AM techniques.

Additionally, when it comes to updating the design, just tell your 3D printer how you want it to look. There's no reason to toss away the money you invested in AM equipment [12].

1.3.2 Simple to Grasp (and Use)

With all new technology and machines, there is a training curve. The good thing is that there is an abundance of readily available instruction on how to utilise AM and 3D printers. There will be suitable training provided, whether you are in charge of the CAD or the operation of the AM equipment [13].

1.3.3 Reduced Waste of Raw Materials

The majority of conventional production processes start with a larger piece of metal before reducing it. Particles of substance that have been milled away frequently have little economic significance. These reductive manufacturing techniques result in significant raw material loss. AM starts from nothing and creates what is needed, reducing the loss of raw materials by up to 90%.

1.3.4 Customisation to an Individual

Since each 3D-produced object depends on electronic data, it is simple to customise each item, eliminating a requirement for retooling. This skill is especially useful in medical and health fields, where customised splints and supports may be personalised to an individual. AM methods are also being researched for the construction of replacement human components, such as bone [14].

1.3.5 Integration of Digital Design

With the advent of CAD, 2D paper sketches gave way to 3D virtual design environments. By going one step further, AM enables those digital 3D drawings to be instantly translated into the real world. Today, a lot of design software supports 3D printing, which makes it easier to automate elements in the design process, like adding inner honeycomb reinforcement or outside scaffolding. Thanks to AM, turning a 3D design into a real-world product is now simpler, quicker, and more affordable than ever [15].

1.3.6 Initial Prototype's Speed

For good reason, the term "rapid prototyping" has been employed to characterise AM. It is simple to develop an initial prototype and assess how it operates in the targeted setting because producing a single item is inexpensive and quick using the AM method. The procedure is able to be repeated after making any necessary modifications or adjustments using CAD. Initial prototypes frequently employ less expensive materials, such as polylactic acid (PLA), whereas later prototypes introduce the materials that are likely to be employed for the finished product.

1.3.7 Reduced Expenses for Energy and the Environment

AM may benefit the environment by cutting down on waste supplies and production time. The LEAP engine's 30,000th printed-in 3D fuel nozzle tip was recently honoured by General Electric. The number of components in a single gasoline nozzle tip was decreased from the approximately 20 pieces that were previously fused together to one complete piece using the AM technique. The weight of the nozzle tip was reduced by around 25%. AM is excellent for the environment and the bottom line.

1.3.8 Low-Production Volume Runs

Low production runs are ideal for AM. Traditional manufacturing processes have higher tooling expenses, which may render low-volume production unprofitable. AM indicates that even one product can be produced profitably.

1.4 CHALLENGES OF AM

The broad use of metal AM is being hindered by a variety of problems. There are drawbacks and obstacles that need more investigation alongside technological development. The limitations on component size, anisotropic mechanical characteristics, the creation of suspended surfaces, high costs, subpar production efficacy, substandard accuracy, buckling, and material usage limits are some of the issues that need further study and analysis. Some of the restrictions and difficulties associated with AM are as follows.

1.4.1 Void Formation

One of the biggest problems of AM components is the porosity creation between succeeding layers. Decreased layer bonding results in this type of issue, which leads to poor mechanical performance. For instance, anisotropic mechanical characteristics are produced by extrusion-based AM methods like FDM, which cause void creation between the manufactured layers. Delamination and, in fact, the kind of AM method and the material employed often affect how much porosity is created by void formation [16].

1.4.2 Stair-Stepping

The occurrence of staircase effects in the manufactured components constitutes one of the most significant obstacles in the AM method. This type of imperfection is minimal for surfaces produced within, but it substantially degrades outside surfaces. Many post-processing methods exist that may mitigate or even eradicate this defect, albeit at the expense of increased production time and material costs [5, 17].

1.4.3 Anisotropic Behaviour

Anisotropy in mechanical characteristics and microstructure is another difficulty that can be seen with AM. By hardening the photosensitive resin, heating the filament, or fusing the powder bed, which produces a heat gradient, AM technologies create components layer by layer. The microstructure and mechanical characteristics of AM components frequently vary depending on the build direction and other directions. For instance, FDM-produced plates are stronger in the x directions than they are in the other directions [17].

1.4.4 Lack of Standards

Standards are crucial in manufacturing because they specify the requirements that must be satisfied to produce a high-quality product. To produce components of the needed quality, raw materials, machinery, equipment, employees, engineers, vendors, and the process of production itself all demand norms and a system for evaluating and certifying against those standards. While the use of 3D printing has been available for over 30 years, it has only lately gained acceptance as a production technique in the manufacturing sector. Because of this, the industry is just now starting to create standards that are applicable to manufacturing. Therefore, one of the primary obstacles preventing 3D printing from becoming more widely used is the absence of standards.

1.5 COMPARISON BETWEEN AM AND SUBTRACTIVE MANUFACTURING

Subtractive manufacturing procedures remove portions of a solid material to make the end components, whereas additive methods build items through the addition of materials layer by layer (Figure 1.3). Although both additive and subtractive

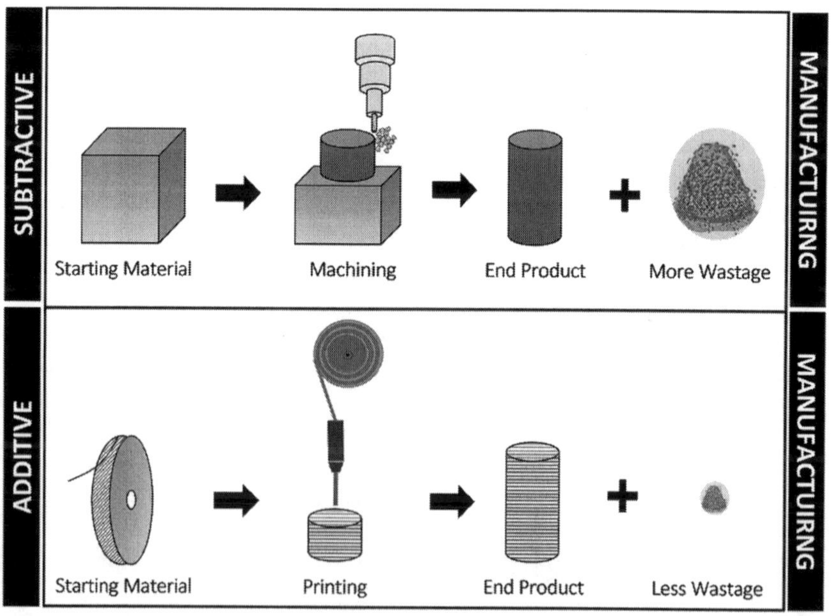

FIGURE 1.3 Schematic representation of process flow followed in additive and subtractive manufacturing.

manufacturing techniques have their own methods for creating prototypes and finished goods, they take different approaches to manufacturing. The following are a few key distinctions between additive and subtractive production.

1.5.1 Material Options

Material selection is a key distinction between additive and subtractive manufacturing when comparing the two processes. The types of materials that are capable of being employed in AM are limited. In addition, synthetic polymers make up the bulk of the supplies utilised during this process. Other materials utilised in the AM process consist of ceramics, biochemicals, thermoplastics, metals, and resins. Subtractive manufacturing, on the other hand, starts with just one component of solid material. Additionally, this substance might be made of glass, wood, plastic, metal, composites, or a derivative of plastic [19].

1.5.2 Complexity

When it applies to complex constructs, AM is the best choice, whereas subtractive manufacturing is most appropriate for simpler designs. However, when it comes to creating complex designs in large numbers, subtractive manufacturing prevails [20].

1.5.3 Accuracy

Subtractive manufacturing equipment is more precise than AM ones. Additionally, machining must follow printing in order to attain high precision when employing AM [21].

1.5.4 Properties of End Product

Surfaces of products created via AM feature very small holes. These holes may additionally result in structural weakness, food contamination, or illness for items utilised in the medical industry. Cleaning and polishing are therefore required to enhance the qualities and look of things created using additive techniques. On the contrary, when products are manufactured using subtractive methods and computer numerical control (CNC) cutting, tougher components are developed. Additionally, these parts are more adaptable and have better finishing than AM-produced components [22].

1.5.5 Surface Finish

Surface finish choice is another thing to take into account when contrasting subtractive manufacturing with AM. When AM is coupled with PBF, the surface quality is subpar. The product's surface is frequently rough due to deposition layers and partly fused powder that are visible [23]. In addition, a product with an uneven surface is more brittle and performs poorly under fatigue. Although they generate a smoother surface, surface machining and shot peening frequently overlook interior surfaces and minute details. On the other hand, CNC subtractive manufacturing results in higher tolerance and nicer surface finishes.

1.5.6 Speed

The pace of an AM process depends on the batch size and design use application. Additionally, one of the benefits of 3D printing over subtractive manufacturing over AM is that it reduces the time needed for small-scale production and speeds up prototyping. However, 3D printing requires more time than subtractive manufacturing to create a single item. In addition, 3D printing takes longer when producing plastic parts in large quantities.

1.6 DIFFERENT TYPES OF AM

AM is one of the methods for transforming complicated 3D models into real-world items. Since it is a layer-by-layer production process, this AM is characterised by the manner of binding (Figure 1.4). AM is categorised into six groups based on the underlying concept of binding.

1.6.1 Material Extrusion

With the help of the AM technique known as material extrusion, a 3D part is developed by layer-by-layer selective deposition of a reel of material (often thermoplastic polymer) that is forced via a hot nozzle in an uninterrupted flow. Material extrusion is often slower and less precise than other kinds of AM techniques. Material extrusion methods and appropriate raw materials, such as nylon and plastic made from ABS, are, on the other hand, widely available and affordable. Therefore, it is the most preferred method for hobbyist-grade 3D printing at home [24]. Nowadays, the researchers are exploiting the use of FDM to develop the metal parts. In such an approach,

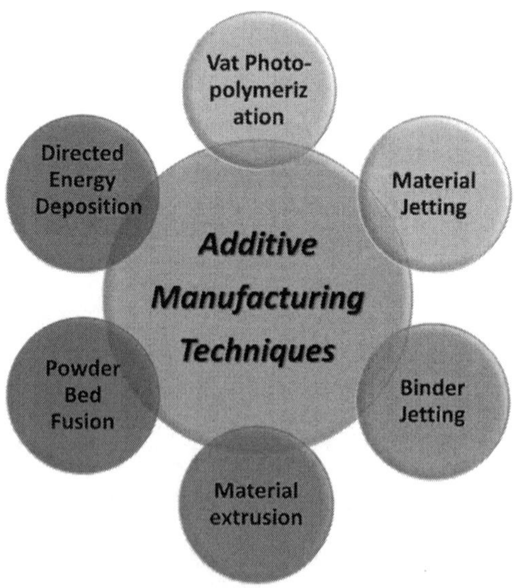

FIGURE 1.4 Different types of AM techniques used to print stand-alone parts.

heating is used to soften the filament so that it may be extruded via a printing nozzle from the feedstock, which consists of metallic powder and polymeric binder. The printed product is then placed on the warmed printing bed, which increases adherence between the produced components and the printing bed, enabling the layer-by-layer construction of the 3D part from the CAD model. The as-printed components need to be de-bonded and fused after printing, similarly to how they are processed during the metal injection moulding process.

1.6.2 Material Jetting (MJ)

MJ is an AM method and is similar to 2D printing in use. In MJ, a printhead (similar to those employed in conventional inkjet printing) dispenses droplets of a photosensitive material that solidifies under ultraviolet radiation, layer-by-layer building a part. The materials used in MJ are liquid thermoset photopolymers (acrylics) that are readily accessible. High-dimensional precision components with an exceptionally smooth finish are produced by MJ 3D printing. It offers multi-material manufacturing and a large selection of materials, including totally transparent and rubber- and ABS-like polymers. These features make MJ a particularly alluring choice for the production of visual prototypes and tools. In one application of the MJ process, drop-on-demand (DOD) printheads are employed to release highly viscous fluids and create objects that mimic wax. However, DOD is almost only used to create investment casting patterns. Therefore, we won't go into greater detail about it here. MJ deposits matter in a line-wise manner, in contrast to the majority of other 3D printing processes. The same carrier is connected to many inkjet printheads placed side by side, which deposit material across the whole printing surface in one pass. This makes it simple and commonplace to print in several materials, in full colour, and to pour dissolvable support structures because separate heads may dispense distinct substances. When printing items with MJ, support structures are perpetually necessary and must be taken out during post-processing. In MJ, the liquid material is made rigid through a process called photopolymerisation. The same approach is used in stereolithography (SLA). Similar to SLA, MJ-printed parts have consistent mechanical and thermal properties, but compared to SLA, MJ-printed components do not require additional post-curing to achieve their optimal qualities because of the incredibly low layer thickness used [25].

1.6.3 Binder Jetting Technique (BJT)

It was put into use in 1999 by ExOne and is based largely on the idea of bonding by adhesives at a specific location obtained via slicing cross-section. A binder substance, which is thought of as the process initiator, controls the formation of coalescence of the particles. Initially, a powder layer is formed, then the glue is selectively poured over the bed utilising a printhead (Figure 1.5). The same is then heated using a separate infrared burner so that the adhesive is partially healed in the locations where the binder has been soaked, and its minimum mechanical characteristics may be obtained. With the help of feed cartridge rollers, an additional layer is once again added after being heated, and the cycle is repeated until the entire required component is created.

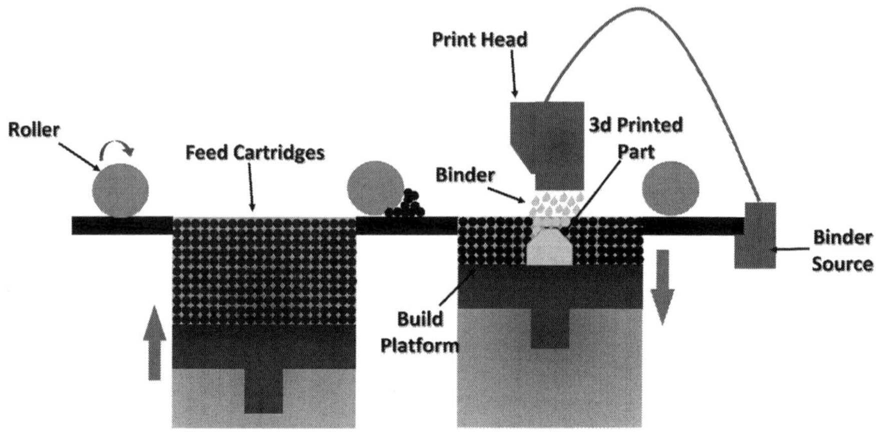

FIGURE 1.5 The schematic representation of the BJT resulting in the 3D printing of standalone parts for end use.

However, the characteristics of the component that is actually created are inadequate for its usage as a finished good, making post-processing procedures like sintering, penetration, finishing, etc., necessary. Additionally, it is fascinating to observe that nearly all of the production time is utilised during additional processing, and little of it is utilised during printing. The entire procedure is comparable to the laser- and electron-based PBF, with the exception of the binder and kind of heat source. Additionally, BJT has a number of advantages over them, including the capacity to process a wide range of material types, avoidance of heat-related issues brought on by insufficient thermal input, and the lack of the need for expensive lasers or electron beams, which makes this approach affordable. Furthermore, since the loose and residual feedstock provides sufficient support for the consolidated component as well as the overhangs, the necessity for support structures is unnecessary. Despite these advantages, BJT also has the distinctive capability of producing coloured items. Its spontaneity might be due to the use of a coloured binder, which gives the portion these qualities. The development of BJT has seen several technical advances since its inception in the late 1990s. Numerous new companies have entered the market in recent years, the majority bringing with them a distinctive viewpoint on this technology [26]. ExOne stands out among them as a leader in the BJT. Over the course of its history, they have added four systems, each one modifying the previous. For instance, ExOne introduced the Innovent+, an upgrade to M-Flex 3D, in 2018. There are two substantial improvements, even if the brand-new machine's time needed for part manufacture is somewhat higher. It has an ultrasonic recoater installed, which is intended to improve material flowability, and it is also simple to switch feedstocks. In the long run, this would provide the BJT the opportunity to build a large chunk of the total production industry [27].

1.6.4 Vat Photopolymerisation (VP)

In the VP technique of AM, photopolymer liquid resin is specially cured using light-activated polymerisation to produce 3D parts (Figure 1.6). A photopolymer is a form of polymer that, when exposed to light, usually radiation, which is within the UV or visible range, transforms the manner in which its molecules respond. It is also known as sunlight-activated resin. Stereolithography (SLA), the first trademarked and widely accessible AM technique, employs the VP approach.

The modern layered stereolithography method was developed by Japanese researcher Dr. Hideo Kodama at the beginning of the 1970s. He used UV light for curing photosensitive polymers. After founding a business named 3D Systems that would market the method and licence it in 1986, Chuck W. Hull created the term SLA. Hull described the method as printing successive small amounts of an ultraviolet-curable material to produce 3D objects.

All VP machines use specialised plastics known as photo polymers as their printing medium. When subjected to specific wavelengths of light, a process known as photopolymerisation leads chemicals in aqueous photo polymers to quickly bond together and consolidate into a solid form. In the vast majority of 3D printers using VP, the platform for the building is frequently partly submerged at the surface of the liquid. The printer carefully regulates the light stream to transform the liquid photopolymer into a rigid layer using information from an Autocad programme. After that, the process is replicated for the subsequent layers up until the entire design is created [28].

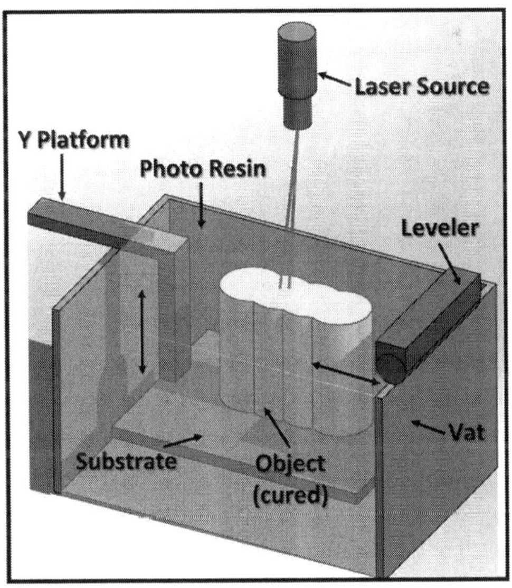

FIGURE 1.6 A simplified illustration of the stereolithography method used to create parts using liquid photosensitive resin.

1.6.5 Directed Energy Deposition (DED)

It is presently used for freeform manufacturing after being frequently used for fixing or changing already-existing components. A concentrated heat source, like a laser, electron beam, electrical arc, etc., is used to create parts. However, using a single heat source is insufficient for the functioning of the system; simultaneous material addition is far more important. Similar to powder bed techniques, the entire procedure takes place while the construct plate is free to travel up or down by a certain amount. The technique is often carried out in a sealed space with an active gas or vacuum to keep careful tabs on the material's characteristics and protect it from oxidation [29]. DED technology has been in use for a long time and offers a number of noteworthy benefits over competing technology. For instance, in contrast to PBF, no feedstock layer develops, but outside addition is accomplished using feeders (Figure 1.7). The inclusion of powder alongside the source of heat also reduces the need for an enormous quantity of feedstock. Additionally, the development of a fracture on a component due to fatigue or any associated problem compromises the structural integrity and necessitates replacement in many applications. To repair the identical portion via material deposition across the required region, however, DED offers a significant benefit and can save a significant amount of money [30]. The DED process uses different heat source methods like laser, arc, and electron beams for consolidation purposes. Figure 1.8 depicts the illustration of the electron beam-based DED process used by the industries.

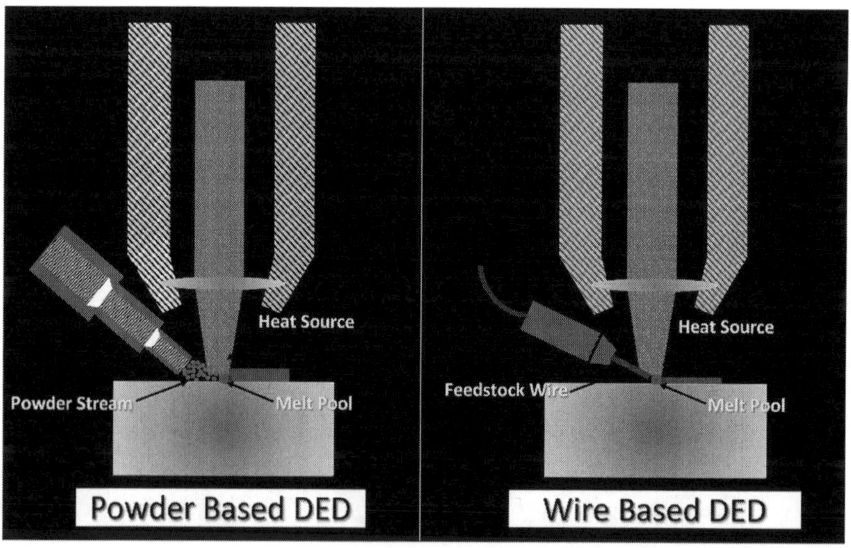

FIGURE 1.7 Schematic illustration of powder and wire-based DED process utilised to 3D print the components.

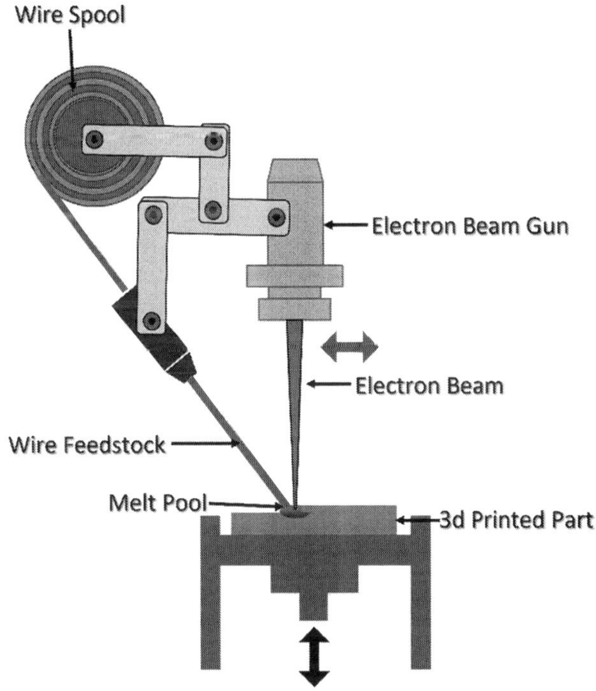

FIGURE 1.8 Diagrammatic representation of the DED method using an electron beam that has been employed for 3D printing the parts.

1.6.6 PBF

The PBF was one of the initial equipment to be integrated into the metal AM production process. Engineers at the University of Texas made efforts in the 1980s that led to the development of this technology [11, 14]. This straightforward approach is changed in at least one way by every other PBF process in order to increase machine productivity, process a variety of goods, and avoid exclusive features. Nevertheless, the core set of characteristics shared by several PBF processes is very uniform. Points are used to illustrate its main mode of activity. This comprises the following:

- The construction platform moves by a certain amount upward or downward.
- The creation of a powder bed over the construction platform using a specific mechanical spreader (such as a roller or surgeon's blade).
- The scanning of a specific area over the bed of powder utilising a heat source to solidify it.
- The cycle repeats until the component is finished.

The type of heat source used is crucial in achieving AM's primary goal for various materials. In some systems, infrared heaters, lasers, and electron beams are frequently used as driving sources and offer the potential for consolidation.

1.6.6.1 Laser-Based PBF

It is recognised as the most widely used thermal source among all those in the market. Such laser-based systems go by a variety of names, including direct metal laser sintering, direct metal laser melting and others. Despite the names being different, each system uses a similar 3D printing methodology. All of these systems choose laser technology because it can deliver enough focused energy to create melt pools in the targeted area. According to the slice data, the galvanometers are responsible for directing the beam of laser light over the area that is relevant in a certain layer. (Figure 1.9) [31]. Additionally, this focused beam is extremely vulnerable to ambient gases and tends to create a molten pool. Such gas mixtures have the potential to result in oxidation or a number of gaseous faults. Such a strategy is thought to be problematic for the printed component. As a result, it is clear that an enclosed space requires a shielding gas to avoid this reactiveness. Nitrogen, helium, argon, and other common gases are employed to achieve these goals. Additionally, only a few ppm of oxygen is needed for the enclosed chamber's procedure to finish successfully [32].

1.6.6.2 Electron Beam-Based PBF

Through the employment of a high-velocity electron beam that concentrates on a specific area, this technology propels the method (Figure 1.10). The most popular pronunciation is "electron beam melting" (EBM). A company called Arcam produced the first system that was sold for profit in 2002 [33]. Compared to other systems, it has excellent process efficacy. Due to the electron's large kinetic energy, it produces the necessary thermal energy for particle aggregation. Yet, the identical technology used in laser PBF is related to the absorption of photons mechanism. Additionally, both lasers and electrons-based PBF have a number of additional fundamental peculiarities. As an example, the printing process is carried out in the required vacuum atmosphere due to the tendency of electrons to get diverted with the contact of atoms. Magnetic coils are also used to regulate the electron beam's velocity. These magnets have the capacity to respond very instantly to changing input circumstances.

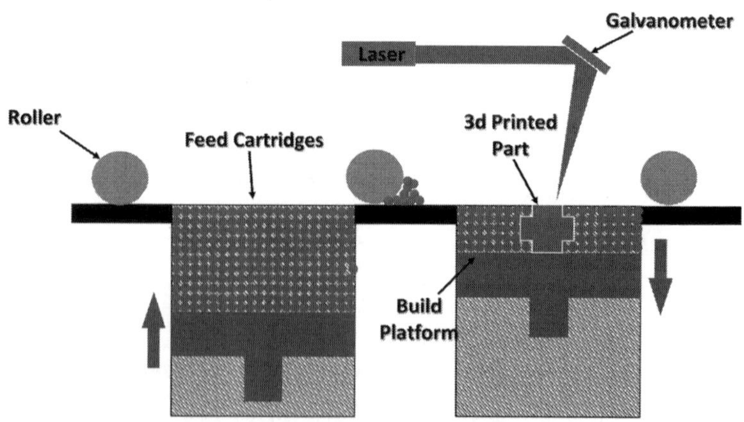

FIGURE 1.9 A demonstration of the laser-based PBF technique used to 3D-print metal parts.

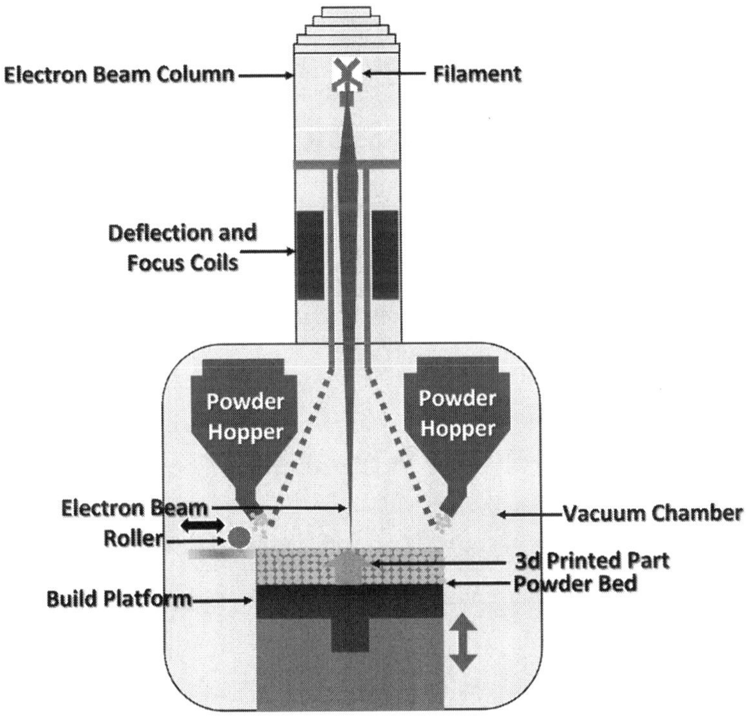

FIGURE 1.10 Schematic representation of electron beam-based PBF process.

As a result, a beam of electrons can be probed slowly or fast. The electron rapid scan essentially refers to the capacity to produce an entire part in a fairly brief period of time. Less lead time, a key characteristic of EBM, denotes great productivity as well as low expenses for every part.

Although we are aware that nothing is flawless, the EBM process is no exception. There are a variety of process skills as well as certain system incompetence. Its inability to treat non-conductive materials is its fundamental limitation [9]. Negative charge collection over a particular region in non-conductive material is obvious and unavoidable. Beam deflection and quick ejection are issues brought on by this charge deposition. Due to these challenges, the material that conducts electricity is ideal for the PBF process based on electrons.

1.7 APPLICATIONS IN VARIOUS SECTORS

Over the last few decades, AM technology has advanced dramatically. Although the technology of AM has been available for quite some time, it has recently gained popularity in terms of different applications. Although new AM applications are continuously being developed, the ones in the following list are currently becoming more significant. The fact that AM is a straightforward innovation that can be applied to a wide range of uses is a significant factor in the recent rise in its utilisation (Figure 1.11). AM was launched with high entry fees in the early years. Models

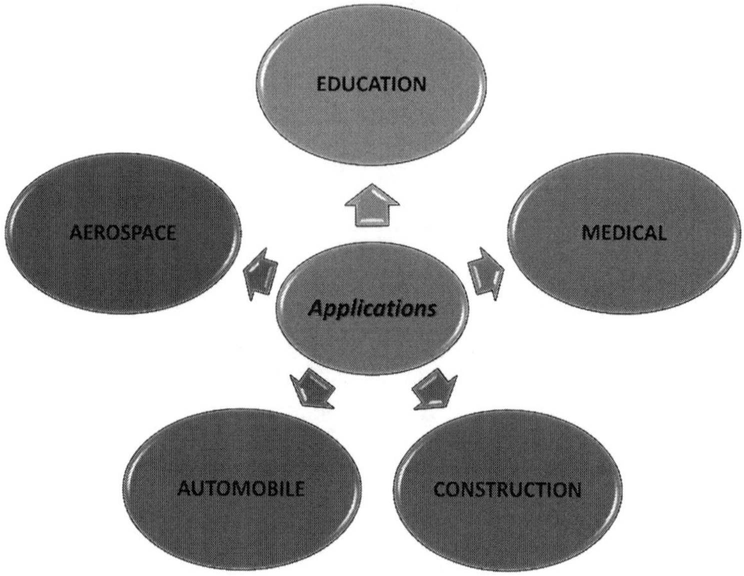

FIGURE 1.11 Chart showing various AM applications.

and materials for AM were expensive. Due to recent advances in both machinery and material developments, costs have decreased, making AM uses more widely available and cost-effective across industries and learning centres [34]. The areas that use AM technology are propelling the globe forward. In that regard, the most essential five applications of AM were covered in this section.

1.7.1 Education

More and more schools incorporate AM technology into their educational programmes. By enabling students to design samples without the use of expensive tools, AM for education can help schools better prepare children for their future careers by designing and producing models that are better suited for trainee apprentices' study about 3D printing capabilities. 3D printing solves any problems associated with ideas and images on a page or computer by allowing those thoughts/pictures to be created in an actual 3D environment. AMs are now frequently seen in educational institutions and libraries. Students can use 3D printers in colleges for assignments and courses. 3D printing devices are also transforming STEM (science, technology, engineering, and mathematics) education by enabling understudies in the classroom to create high-quality scientific ideas with little effort using open hardware blueprints. AM assists in learning about a variety of AM applications by examining configuration, layout, and engineering norms. They can view geological markers and charts from fresh, 3D angles. Visual design students may undoubtedly produce models with complex practical components. Science students have the ability to make and study cross-sections of bodily parts. Understudies in science can create 3D representations of atoms and chemical substances. Also, all potential notions of practical instruction address problem-solving with creative design [35].

1.7.2 Medical

Medical personnel may now give a 3D image of the field thanks to AM technology. The method has been utilised (and continues to be used) by a number of businesses, especially those that focus on therapeutic innovation. The original advanced model, which is then produced in the 3D printer, is frequently created using ultrasounds, X- Rays computed tomography (CT) scans, and magnetic resonance imaging (MRI) results. Since these organoids imitate tissues on a smaller scale than usual, they can be employed in biomedical investigation. They are additionally being researched as less costly substitutes for organ donations in humans. The construction of patient-specific organ models that professionals can use to practise before performing challenging operations is another use for 3D printing in health. It has been demonstrated that using this method accelerates psychological harm in people. Forceps, hemostats, and bracing are examples of clean medical instruments that can be created using 3D printers. Additionally, AM enables the patient to construct a prosthetic that is uniquely suited to their requirements. Body Labs, for example, has developed a structure that enables patients to properly build a connection between their prosthesis and their leg. A few other medicinal applications of AM include cosmetic procedures and skincare items. During the COVID-19 epidemic, the AM was critical in resolving concerns with N-95 masks [36]. Because of the presence of pollutants in the used filter, it is generally accepted that such masks must be discarded after a single usage. ExOne, a firm based in 2020, began investigating the use of BJT to make replaceable copper filters in order to substitute throwaway N-95 masks. ExOne intended to create copper water purifiers but instead came up with a coronavirus-related innovation. Copper is extremely difficult to manufacture into a thick component due to its poor optical properties. However, it was believed that this flaw would be advantageous in the creation of copper filters with a highly controlled permeability. These tiny pores have an endless bulk path and are porous in dimension. Its ability to capture viruses in these openings and ultimately eradicate them makes its use as a filter as well.

1.7.3 Construction

The idea of rethinking our approach to modern building through the use of AM technology is an exciting but cautious study subject. Despite the fast acceptance of AM in a variety of industries, including electronics, medicinal, defence, and aerospace, accessible building projects are primarily in the planning stages. To deal with the restricted availability of specialised labour and to undertake sensitive building processes, progressive nations are intensively investigating the prospect of mechanised construction techniques [37]. AM, or 3D printing, with its layer-upon-layer production capacity, offers special benefits in construction, such as simplifying beam links, optimised distribution of stresses in structures, decreased duration of construction, effective energy consumption, decreased waste of materials, improved security during development, and effortless connection of printing settings with building information modelling (BIM). Unlike traditional building approaches, which almost usually result in increasing expenditures for sophisticated structures, the enhanced intricacy and mechanisation do not need additional expenses [38]. Without the constraints of traditional building geometries, structural sections may be correctly modelled and positioned in appropriate positions. Material utilisation can also be economical without sacrificing load-bearing capability or other essential features. Because the construction sector has a significant environmental effect (it accounts for 36% of yearly worldwide energy usage and 39% of power-related greenhouse gas emissions [8]), adopting AM technology can also help with carbon mitigation and give other ecological advantages [9]. Modifications in design may be accepted later in the process of building, and components can be fabricated reliably on or off-site. Laser AM techniques may also be used for the maintenance and repair of damaged metallic parts. Certain AM approaches are also being investigated for the quick repair of shelters in places hit by unanticipated disasters [39]. Furthermore, because of the quick cooling speeds intrinsic in the printing by a laser process, AM can achieve distinct mechanical characteristics for the same material. The use of heat and thermally induced prestressing are two post-processing procedures that can further customise the obtained material characteristics. In the meantime, the off-site building has grown in favour over the last few decades. Off-site construction encompasses an extensive variety of construction projects, each with its own definition. Off-site building, by definition, refers to a construction method in which building parts are created in areas other than their eventual deployment locations. This has been referred to variously as panelised building, prefabrication, modular construction, and so on [40]. Two tactics have been utilised in its progress over the last decades: product and production concepts. Building elements are viewed as goods that may be standardised for prefabrication in these proposals. This has resulted in increased production and increased popularity in the building business.

1.7.4 Automobile

To stay relevant in the era of Industry 4.0, automakers are using AM technologies more and more. AM helps manufacturers remain flexible and creative at all stages of the car business, from trials and manufacturing of tools to completed products. Furthermore, automakers use metal AM to effectively generate superior components with complex shapes and structures. In the automotive industry, AM is frequently used to make parts for the powertrain, the chassis, the suspension, the seat structures, the exhaust cooling systems, and other components. AM is regarded as capable of creating the complete 3D-printed automobile for use in addition to manufacturing parts. They use lightweight materials in one of their AM processes, known as FDM [41]. Moreover, one of the well-known enterprises, Bugatti, 3D-manufactured titanium braking calipers to decelerate the automobile from 260 mph to rest [42]. However, the handling of conventionally made calipers was significantly impacted by their extensive utilisation. In order to maintain the driving characteristics, Bugatti developed the calipers using the metallic AM technique, which led to a 40% weight reduction over the component made conventionally. Furthermore, the ability of metal AM to manufacture interior lattice structures has enabled the vehicle sectors to reduce material and money waste. Furthermore, the fabrication of such parts results in weight reduction, which is attributed to better energy utilisation. For instance, the business Ceramic Disc Technologies (CDT) used the same mentioned lattice structure to construct the brake discs for the Challenger, preserving a significant amount of resources in the process [11]. Yet, when it comes to reducing weight, this AM-fabricated item provides significant variety. Furthermore, the existence of a lattice structure allows for optimal air movement, resulting in greater heat transmission. AM has made automotive-related organisations more adaptable, allowing them to create better automobile components that would have been impossible to create using traditional methods. However, along with the benefits of the procedure, there are also inherent limitations. The biggest problem is that the machine cannot produce large numbers of vehicle components. As a consequence, AM in cars is likely to be used for bulk production of parts to satisfy consumer demand.

1.7.5 Aerospace

AM was first developed as a manufacturing process to repair damaged components in aircraft, consequently saving large amounts of money [43]. However, as technology advances, aerospace expands the use of additive printing beyond repair technologies to independent part fabrication. With this broadening of implementation, the AM provided various benefits. For example, reducing the amount of parts in an assembly results in cost savings in terms of manufacturing and assessment [44]. GE Aviation reported a decrease from 855 pieces produced using traditional methods to just a few components produced using AM methods. As a result, part consolidation saves weight and improves component effectiveness. Furthermore, the technology allows developers to create and manufacture rocket elements in a shorter amount of time. The ArianeGroup, in particular, created an injector mechanism out of nickel-based alloy utilising an EOS M 400-4 metal AM machine [45]. The same product formerly took over three months to fabricate using casting and machining, but AM reduced the processing time to 35 hours. Other essential aerospace-related uses include the fabrication of helicopter combustion chambers (aerospace applications book), airfoil repair, quick tooling to construct Aircraft A320 hydraulic hinges for the doors, and so on [46]. Masten Space Systems has made recent developments in aerospace applications. They created a lightweight, durable missile engine to help NASA achieve its objective of making multiple trips to the moon in fast succession with minimal intervals between flights. It was referred to as a "Broadsword" engine created using the PBF process – namely, DMLS by EOS. Furthermore, AM resulted in the formation of sophisticated air-cooled pathways inside the main body of Broadsword's combustion chamber and the unification of the engine's parts into only three major pieces. Furthermore, in order to meet future production demands, metal AM must overcome significant challenges. Among these include AM part validation, better process repeatability, and safety. Nonetheless, with major investments already being spent in developing and validating AM equipment and feedstock, the prospects for the use of AM in the aerospace industry are bright.

REFERENCES

[1] D. W. Rosen and B. S. Gibson, "Generalized additive manufacturing process chain," in *Additive Manufacturing Technologies*, Cham, Switzerland: Springer US, 2015, pp. 43–61.

[2] C. W. Hull, "Apparatus for production of three-dimensonal objects by stereolithography," 1984 [Online]. Available: US Patent 4,575,330.

[3] Stratasys, "The history of stratasys." https://www.stratasys.co.in/ (accessed Apr. 20, 2021).

[4] EOS GmbH, "The eos story." https://www.eos.info/en/about-us/history (accessed Apr. 20, 2021).

[5] I. Gibson, D. W. Rosen, and B. Stucker, "Powder bed fusion processes," in *Additive Manufacturing Technologies: Rapid Prototyping to Direct Digital Manufacturing*, Boston, MA: Springer US, 2010, pp. 120–159.

[6] W. E. Frazier, "Metal additive manufacturing: A review," *Journal of Materials Engineering and Performance*, vol. 23, no. 6, pp. 1917–1928, 2014.

[7] Autodesk, "Tinkercad." https://www.tinkercad.com/ (accessed Apr. 25, 2021).

[8] V. T. Le et al., "Prediction and optimization of processing parameters in wire and arc-based additively manufacturing of 316L stainless steel," *Journal of the Brazilian Society of Mechanical Sciences and Engineering*, vol. 44, no. 9, Sep. 2022, doi: 10.1007/s40430-022-03698-2

[9] B. Cheng and K. Chou, "Thermal stresses associated with part overhang geometry in electron beam additive manufacturing: Process parameter effects," *25th Annual International Solid Freeform Fabrication Symposium &#65533; An Additive Manufacturing Conference, SFF 2014*, pp. 1076–1087, 2014.

[10] J. Jiang, X. Xu, and J. Stringer, "Support structures for additive manufacturing: A review," *Journal Of Manufacturing And Materials Processing*, vol. 2, no. 4, p. 64, 2018.

[11] A. Singh and H. Singh, "Metal additive manufacturing: From history to applications," in *Innovations in Additive Manufacturing*, M. A. Khan and J. T. W. Jappes, Eds., Cham, Switzerland: Springer International Publishing, 2022, pp. 3–32, doi: 10.1007/978-3-030-89401-6_1

[12] GE Additives, "Metal additive manufacturing materials." n.d.

[13] J. Karthikeyan, "The advantages and disadvantages of the cold spray coating process," in *The Cold Spray Materials Deposition Process*, V. K. Champagne, Ed., Woodhead Publishing Series in Metals and Surface Engineering, Cambridge, MA: Woodhead Publishing, 2007, pp. 62–71.

[14] A. Singh, P. Singh, B. S. Pabla, H. Singh, and S. Shiva, "Parametric analysis to explore the viability of cold spray additive manufacturing to print SS316L parts for biomedical application," *Journal of the Brazilian Society of Mechanical Sciences and Engineering*, vol. 44, no. 8, p. 339, Aug. 2022, doi: 10.1007/s40430-022-03666-w

[15] A. Vargas-Uscategui, P. C. King, S. Yang, C. Chu, and J. Li, "Toolpath planning for cold spray additively manufactured titanium walls and corners: Effect on geometry and porosity," *Journal of Materials Processing Technology*, vol. 298, p. 117272, Dec. 2021, doi: 10.1016/j.jmatprotec.2021.117272

[16] S. Tammas-Williams, P. J. Withers, I. Todd, and P. B. Prangnell, "Porosity regrowth during heat treatment of hot isostatically pressed additively manufactured titanium components," *Scripta Materialia*, vol. 122, pp. 72–76, 2016, doi: 10.1016/j.scriptamat.2016.05.002

[17] S. Yin, R. Jenkins, X. Yan, and R. Lupoi, "Microstructure and mechanical anisotropy of additively manufactured cold spray copper deposits," *Materials Science and Engineering A*, vol. 734, no. January, pp. 67–76, 2018, doi: 10.1016/j.msea.2018.07.096

[18] V. Lindström et al., "Laser powder bed fusion of metal coated copper powders," *Materials (Basel)*, vol. 13, no. 16, p. 3493, Aug. 2020.

[19] M. Moussa and H. ElMaraghy, "Multiple platforms design and product family process planning for combined additive and subtractive manufacturing," *Journal of Manufacturing Systems*, vol. 61, pp. 509–529, Oct. 2021, doi. 10.1016/j.jmsy.2021.09.019

[20] S. A. Khairallah, A. T. Anderson, A. Rubenchik, and W. E. King, "Laser powder-bed fusion additive manufacturing: Physics of complex melt flow and formation mechanisms of pores, spatter, and denudation zones," *Acta Materialia*, vol. 108, pp. 36–45, 2016, doi: 10.1016/j.actamat.2016.02.014

[21] Z. Wang, S. Zimmer-Chevret, F. Léonard, and G. Abba, "Improvement strategy for the geometric accuracy of bead's beginning and end parts in wire-arc additive manufacturing (WAAM)," *International Journal of Advanced Manufacturing Technology*, vol. 118, no. 7–8, pp. 2139–2151, Feb. 2022, doi: 10.1007/s00170-021-08037-8

[22] J. Li, L. Cheng, P. Zhang, L. Wang, and H. Li, "Effect of delta ferrites on the anisotropy of impact toughness in martensitic heat-resistant steel," *Journal of Materials Research and Technology*, vol. 8, no. 2, pp. 1781–1788, Apr. 2019, doi: 10.1016/j.jmrt.2018.12.008

[23] Y. H. P. Manurung, K. P. Prajadhiana, M. S. Adenan, B. Awiszus, M. Graf, and A. Haelsig, "Analysis of material property models on WAAM distortion using nonlinear numerical computation and experimental verification with P-GMAW," *Archives of Civil and Mechanical Engineering*, vol. 21, no. 1, p. 32, Mar. 2021, doi: 10.1007/s43452-021-00189-4

[24] G. Shayegan et al., "Residual stress induced by cold spray coating of magnesium AZ31B extrusion," *Materials and Design*, vol. 60, pp. 72–84, Aug. 2014, doi: 10.1016/j.matdes.2014.03.054

[25] O. Gülcan, K. Günaydın, and A. Tamer, "The state of the art of material jetting—A critical review," *Polymers*, vol. 13, no. 16. MDPI AG, Aug. 2, 2021, doi: 10.3390/polym13162829

[26] H. Miyanaji, "Binder jetting additive manufacturing process fundamentals and the resultant influences on part quality," 2018.

[27] M. Leary, "Binder jetting," in *Design for Additive Manufacturing*, M. Leary, Ed., Additive Manufacturing Materials and Technologies, Amsterdam, Netherlands: Elsevier, 2020, pp. 335–339.

[28] K. Salonitis, "Stereolithography," in *Comprehensive Materials Processing*, Oxford: Elsevier, 2014, pp. 19–67, doi: 10.1016/B978-0-08-096532-1.01001-3

[29] D.-G. Ahn, "Directed energy deposition (DED) process: State of the art," *International Journal of Precision Engineering and Manufacturing – Green Technology*, vol. 8, no. 2, pp. 703–742, 2021.

[30] S. Ehmsen, L. Yi, and J. C. Aurich, "Process chain analysis of directed energy deposition: Energy flows and their influencing factors," *Procedia CIRP*, vol. 98, pp. 607–612, 2021.

[31] T. Taufek, Y. H. P. Manurung, S. Lüder, M. Graf, and F. M. Salleh, "Distortion analysis of SLM product of SS316L using inherent strain method," *IOP Conference Series: Materials Science and Engineering*, vol. 834, no. 1, p. 012011, 2020, doi: 10.1088/1757-899X/834/1/012011

[32] P. Lassègue et al., "Laser powder bed fusion (L-PBF) of Cu and CuCrZr parts: Influence of an absorptive physical vapor deposition (PVD) coating on the printing process," *Additive Manufacturing*, vol. 39, p. 101888, 2021.

[33] L. E. Murr and S. M. Gaytan, "Electron beam melting," in *Comprehensive Materials Processing*, Oxford: Elsevier, 2014, pp. 135–161.

[34] A. Aimar, A. Palermo, and B. Innocenti, "The role of 3d printing in medical applications: A state of the art," *Journal of Healthcare Engineering*, vol. 2019, p. 5340616, 2019.

[35] E. Novak, "3D printing in education," in *3D Printing in Education*, Routledge, 2022, doi: 10.4324/9781138609877-REE81-1
[36] R. Elisheva, "Adverse effects of prolonged mask use among healthcare professionals during covid-19," *Journal of Infectious Diseases and Epidemiology*, vol. 6, no. 3, pp. 6–10, 2020.
[37] S. El-Sayegh, L. Romdhane, and S. Manjikian, "A critical review of 3D printing in construction: Benefits, challenges, and risks," *Archives of Civil and Mechanical Engineering*, vol. 20, no. 2, p. 34, Jun. 2020, doi: 10.1007/s43452-020-00038-w
[38] Y. W. D. Tay, B. Panda, S. C. Paul, N. A. Noor Mohamed, M. J. Tan, and K. F. Leong, "3D printing trends in building and construction industry: A review," *Virtual and Physical Prototyping*, vol. 12, no. 3, pp. 261–276, Jul. 2017, doi: 10.1080/17452759.2017.1326724
[39] W. Li, K. Yang, S. Yin, X. Yang, Y. Xu, and R. Lupoi, "Solid-state additive manufacturing and repairing by cold spraying: A review," *Journal of Materials Science and Technology*, vol. 34, no. 3, pp. 440–457, 2018, doi: 10.1016/j.jmst.2017.09.015
[40] K. Arunprasath, V. Arumugaprabu, P. Amuthakkannan, R. Deepak Joel Johnson, and S. Vigneshwaran, "Development in additive manufacturing techniques," in *Innovations in Additive Manufacturing*, M. A. Khan, J. T. W. Jappes, Eds., Cham, Switzerland: Springer International Publishing, 2022, pp. 33–53, doi: 10.1007/978-3-030-89401-6_2
[41] Stratasys, "The first fuel-efficient 3d printed car is back on the map." https://www.stratasysdirect.com/industries/transportation/3d-printed-car-fuel-efficient-fdm-urbee-2 (accessed May 26, 2021).
[42] Bugatti, "World premier – brake caliper from 3d printer." https://www.bugatti.com/media/news/2018/world-premiere-brake-caliper-from-3-d-printer/ (accessed May 26, 2021).
[43] R. Liu, Z. Wang, T. Sparks, F. Liou, and J. Newkirk, "Aerospace applications of laser additive manufacturing," in *Laser Additive Manufacturing*, M. Brandt, Ed., Woodhead Publishing Series in Electronic and Optical Materials, Cambridge, MA: Woodhead Publishing, 2017, pp. 351–371.
[44] T. Duda and L. V. Raghavan, "3D metal printing technology," *IFAC-PapersOnLine*, vol. 49, no. 29, pp. 103–110, 2016.
[45] ArianeGroup, "Future ariane propulsion module simplified." https://www.eos.info/en/3d-printing-examples-applications/all-3d-printing-applications/aerospace-additive-manufacturing-for-ariane-injection-nozzles (accessed May 22, 2021).
[46] J. C. Najmon, S. Raeisi, and A. Tovar, "Review of additive manufacturing technologies and applications in the aerospace industry," in *Additive Manufacturing for the Aerospace Industry*, F. Froes, R. Boyer, Eds, Amsterdam, Netherlands: Elsevier Inc., 2019, pp. 7–31.

2 Wire Arc Additive Manufacturing

Process Overview and a Case Study on Low-Carbon Steels

Van Thao Le and Tat Khoa Doan
Le Quy Don Technical University, Hanoi, Vietnam

2.1 INTRODUCTION

Additively manufacturing (AM) appeared in the 1980s with stereolithography – a process of solidifying liquid polymer by laser. AM consists of bonding material to form solid physical parts according to the layer-by-layer deposition principle [1–3]. Due to this principle, AM becomes an excellent solution for making complex shapes or components of expensive and/or difficult-to-cut materials (e.g., titanium and nickel alloys). AM technology also allows the reduction of scrap and environmental pollution impacts as compared to conventional manufacturing technologies such as casting, forging, and machining [4–8], especially in cases where a considerable quantity of chips must be machined to achieve the final parts by machining.

Recently, the research community on AM focused on investigating the technology and used materials. AM technologies, especially metallic AM, are applied in rapid prototyping and manufacturing products for direct utilisation, for example, in the automobile and aerospace industries [9, 10]. According to the AMPOWER report in 2019,[1] the global AM technology market reached a value of 2.02 billion euros (including equipment, raw materials, and services). It continued to grow at 27.9% for suppliers and 23.8% for buyers in 2024, with the leading market share focusing on AM technology using powder and wire. The impressive development of AM technologies is now proving to be a breakthrough in the manufacturing technology industry and simultaneously opening the way to optimise product design. Compared to subtractive manufacturing processes, AM is considered an environmentally friendly manufacturing technology [4, 11, 12] because AM creates a shape close to the final product by adding material in layers without using cutting tools and cutting fluids.

DOI: 10.1201/9781003363415-2

Based on the energy source used, there are three groups of metal AMs, as follows [13, 14]:

- laser-based metallic AM technologies,
- AM technologies using the energy of an electron beam, and
- AM technology using an arc welding source.

Laser-based metallic AM, for example, direct metallic deposition (DMD) and selective laser melting (SLM) [14–17], enable high precision fabrication and low surface roughness. However, these processes have low energy efficiency and a long production time. Meanwhile, electron beam melting (EBM) [1, 14–19] has higher energy efficiency. Nevertheless, this process requires an environment of high vacuum, thus limiting its volume fabrication. SLM and EBM are only effective for fabricating small and medium parts.

The DMD process can produce larger components as compared to SLM and EBM. However, DMD presents a low material usage efficiency. In contrast, wire and arc welding-based AM (WAAM) technologies are promising for manufacturing large-scale metallic components with low production costs [20]. The use of fibrous materials is cheaper and safer for operators.

Focusing on the WAAM technology, this chapter first presents an overview of this technique in terms of the principle, classification, materials, and applications. Second, the WAAM process of steels is introduced with a focalisation on the metallurgical characterisation and mechanical properties of as-built components.

2.2 OVERVIEW OF WIRE ARC ADDITIVE MANUFACTURING

2.2.1 Definition

As mentioned earlier, WAAM is a metallic AM process in which an arc is utilised as the energy source to melt and deposit the metallic wire into the part layer by layer. The movement of making the product shape is performed by a robot, a multi-axis computer numerical control (CNC) machine (Figure 2.1). During the deposition process, protective gases (such as argon and CO_2) are used to avoid the oxidation of molten materials from the environment [22].

2.2.2 Classification

Based on the heat source used, WAAM can be divided into three groups (Figure 2.2) – namely, (i) WAAM based on GMAW (gas metal arc welding) source with a consumable electrode (i.e., metallic wire), (ii) WAAM using a gas tungsten arc welding (GTAW) with non-consumable electrodes, and (iii) WAAM using a plasma arc welding (PAW) source [21, 23, 24].

In GMAW-WAAM, the electric arc is formed between the consumed metallic wire and the workpiece. The melted metal is dropped in the welding pool. There are four methods of transferring melted metals, including short circuit spray, globular, and pulsed spray. In contrast, in GTAW- and PAW-WAAM, the tungsten electrode is

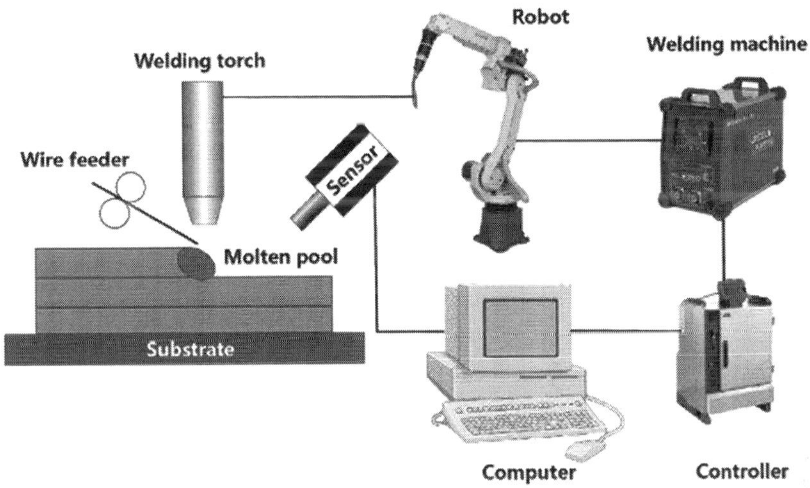

FIGURE 2.1 A WAAM system using gas tungsten arc welding [21].

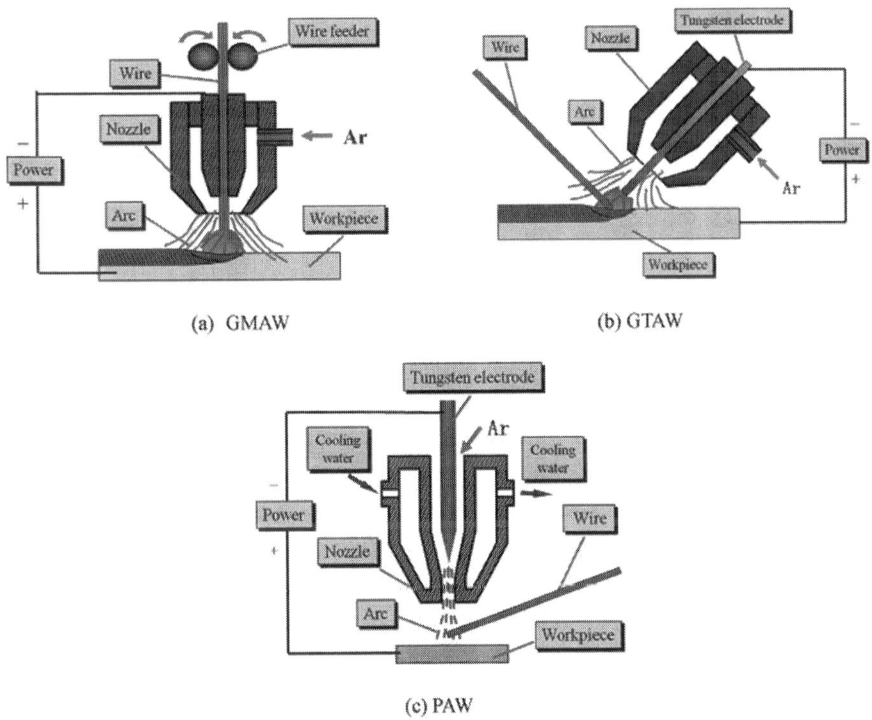

FIGURE 2.2 Arc sources used in WAAM: (a) GMAW, (b) GTAW, and (c) PAW [21].

TABLE 2.1
Different WAAM Technologies and Their Characteristics [25, 26]

Energy Source	Main Characteristics
GMAW	– An electrical arc is formed between the metallic wire and the workpiece. Thus, it is less stable. – There are three methods: short circuit spray, globular, and pulsed spray. – Depositing rate: 3–4 kg/h.
GTAW	– Rear, side, and front feeding. The front feeding is generally used for the WAAM process of titanium and steel alloys. – The electrodes are not consumed, independent feeding, and the average deposition rate is 1–2 kg/h.
PAW	– Non-consumable electrodes, independent of the wire feeding, average deposition rate: 2–4 kg/h. – Manufacture parts with thick or thin wall structures with proper feeding position.

not melted. The filler material can be supplied into the welding area with different orientations, thus influencing the stability of deposition and making the depositing path planning more difficult [25].

Table 2.1 reveals the characteristics of WAAM technologies. The GMAW-WAAM has a higher depositing rate than those in PAW- and GTAW-WAAM. Nevertheless, as the current acts on the welding wire directly, the stability of the GMAW-WAAM process is lower, generating more spatters and smoke than the GTAW- and PAW-WAAM processes. The choice of the WAAM technologies directly influences process conditions for a fabricated part, efficiency, and productivity. Compared to DED technologies, the WAAM's energy efficiency is higher, has low costs, and uses various metals such as titanium-, aluminium-, and nickel-based alloys and steels [25].

2.2.3 Advantages and Disadvantages of WAAM Technologies

WAAM is an emerging AM technology with great interest in research and industrial applications. The WAAM technologies feature the following remarkable advantages [25, 26]:

- The equipment is not too complicated and can be easily implemented by welding robots or integrating a welding source into a CNC machine. Welding wire materials are commercially available. This technology allows the fabrication of thin-walled parts with large dimensions ~ (1,000–3,000) mm, with very elevated deposition rates, ~ (50–130) g/min [24, 27, 28].
- The use of wire material also brings greater material usage efficiency (~ 90–99%). The wire is also more friendly to operators and reduces negative environmental impact. Compared to conventional manufacturing processes, WAAM systems can reduce manufacturing time by 40%–60% [29].

However, this technology also has several disadvantages that need to be studied and improved [30]:

- Thermal distortions and residual stresses caused by a high heat input lead to remarkable deviations in the shape and dimensions of as-built components.
- Low dimensional accuracy and elevated surface roughness/waviness. Thus, functional surfaces of the as-built components must be machined to meet the designed specifications.
- The WAAM components undergo highly complex thermal cycles, significantly influencing microstructures, grain size, and mechanical properties.

These challenges have posed many research problems for scientists in optimising technological parameters, improving the process quality, and expanding the WAAM's applications to various fields.

2.2.4 Materials

The materials used in WAAM are commonly available in the market. They have been studied and fabricated by WAAM for many applications in different industries (Table 2.2).

Steels. They are the most studied with the WAAM technology because the steel welding wires are ready in the market and have good weldability. Steels also have many applications in aeronautics, automotive, tooling, shipbuilding, construction, and so on.

Titanium alloys. These alloys are largely utilised in the aeronautic industry thanks to their lightweight and high-strength characteristics. In addition, WAAM is growingly accepted to manufacture large-scale aerospace parts to save manufacturing and materials costs. Among these alloys, Ti6Al4V is the most popularly investigated with the WAAM process. The tensile strengths of as-built Ti6Al4V components are generally close to those of wrought Ti6Al4V and cast Ti6Al4V [26]. Additionally, Ti6Al4V components built by WAAM processes reveal anisotropic characteristics. This is due to the α-lamella grain size and the prior β grains' orientation.

TABLE 2.2
Commonly Used Materials of WAAM in Industry [26]

Application Industry	Materials
Aerospace	Titanium, aluminium, and nickel-based alloys
Automotive	Steel and aluminium alloys; bimetal
Marine	Steel and titanium alloys
Corrosion resistance	Titanium and nickel-based alloys; bimetal
Elevated temperature	Titanium and nickel-based alloys; bimetal
Tools and moulds	Steels

Nickel alloys. Nickel-based alloys are the most utilised in the aviation, petrochemical, mining, and marine industries. They feature excellent strengths and oxidation resistance at high temperatures (≥550°C). Among these alloys, Inconel 718 and Inconel 625 are commonly investigated and fabricated with the WAAM processes. The microstructures of WAAMed Inconel 718 components are mainly composed of columnar grains with small phases of laves and carbides of MC. On the other hand, the microstructures of WAAMed Inconel 625 show columnar dendrite morphologies, which are decorated with MC carbides, Ni_3Nb, and Lave phases. The mechanical properties of WAAMed Inconel 718 alloys generally fall between the value range of cast and wrought materials. At the same time, elongation is much inferior to that of cast and wrought components. The tensile strengths of Inconel 625 components built by WAAM also meet requests for cast materials but are lower than those of the wrought materials.

Aluminium alloys. They are suitable for products with a relatively complex and thin-walled structure with low manufacturing costs for industrial applications. However, the most crucial issue related to the WAAM process of aluminium alloys (particularly Al 6xxx and Al 7xxx) is the control of melting, solidification, and defects [31]. The WAAMed components made of aluminium alloys often exhibit lower mechanical properties than those of components fabricated by conventional manufacturing technologies – e.g., casting and machining. Therefore, after the WAAM process, the as-built aluminium alloy components must undergo heat treatment to enhance mechanical properties.

Other metals. They include magnesium alloys, for example, AZ31, which are used in automotive industries. Fe/Al and Al/It alloys, bimetallic/Ni steels, and steel/copper are also studied and developed for aerospace applications.

2.2.5 Defects of the WAAMed Components

The quality and accuracy of parts manufactured by the WAAM technologies are highly dependent on technological parameters, including the current, travel speed, voltage, wire feed speed, and heat input [25, 26, 32]. During the WAAM process, the layer temperature is continuously increased because the heat from the newly deposited layer results in the localised heat build-up phenomenon, reducing the dimensional and surface accuracy. The heat build-up between layers can also cause reduced material properties, making the material brittle and easy-to-produce defects.

Currently, the components fabricated by WAAM exhibit many defects, which are due to improperly selected process conditions, considerable heat accumulation, environmental effects, and other problems related to the equipment. Common defects occurring in the WAAMed products include the following [26, 33]:

Distortion and residual stresses. Residual stress is the leading cause of deformation, making the geometrical shape deviations and reducing mechanical properties. Residual stresses are due to large temperature differences in the melting and rapid cooling processes.

Porosity. This defect needs to be mitigated during the WAAM process. Porosity influences fatigue strengths and mechanical resistance. It is also susceptible to microcellular cracks in material surfaces. Typically, pores are formed in weld beads because of the unclean workpiece surface and the weld wire material contaminated with impurities. The pollutants will be absorbed as the material melts and forms after solidification.

Delamination and cracks. These defects relate to thermal cycles in WAAM and material characteristics. Cracks existing in as-WAAMed parts can be classified into solidified cracks or cracks at grain boundaries.

Typically, heat input in WAAM is favoured by the conduction through the substrate, the convection of the protective gas flow, and the radiation to the air. Therefore, controlling heat input can improve the product shape and mechanical characteristics and reduce residual stresses. Currently, various methods have been applied to cool down and keep the stable temperature between layers, as follows:

First, it is possible to limit the heat build-up by adding an idle time between two consecutive deposited layers [34]. In this case, it is possible to combine simultaneously with temperature monitoring to ensure that the interlayer temperature between the layers is similar. The interlayer temperature augments with the deposition height. Consequently, the applied idle time must be increased to maintain a similar interlayer temperature [35]. However, increasing the idle time leads to an increase in the total building times, thus decreasing productivity.

Second, external cooling methods can be utilised to minimise the phenomenon of heat accumulation. In this case, the building table can be immersed in a water tank. Thus, the heat is transferred to the water environment directly or through the substrate [36, 37]. However, the disadvantage of this method is that the arc is still overflowing to both sides. The cooling rate is also not guaranteed continuously during the process. This method is also relatively cumbersome or even a bit complicated to implement, therefore reducing the flexibility of WAAM.

Third, integrating a compressed air-cooling system using inert gases, such as CO_2, N_2, and Ar, into the WAAM system [31, 38–43]. This method enables reducing the heat build-up in the top deposited layers [23, 44]. The compressed air-cooling system reveals better efficiency while reducing costs because of its convenience and availability.

2.2.6 Applications

WAAM is increasingly applied in aerospace, automotive, shipbuilding, defences, tooling, and nuclear energy. In the shipbuilding industry, typical products such as propellers, lifting assemblies, and manipulators can be fabricated by the WAAM process.

In the aerospace industry, the products mainly focus on parts made from it and Ni alloys. In this case, the WAAM process shows superiority over conventional manufacturing technologies, e.g., forging and machining. Besides, WAAM allows an increasing ability to optimise the design.

Figures 2.3 and 2.4 show examples of components in various shapes, from simple to complex, produced by the WAAM technology. Therefore, this technology is

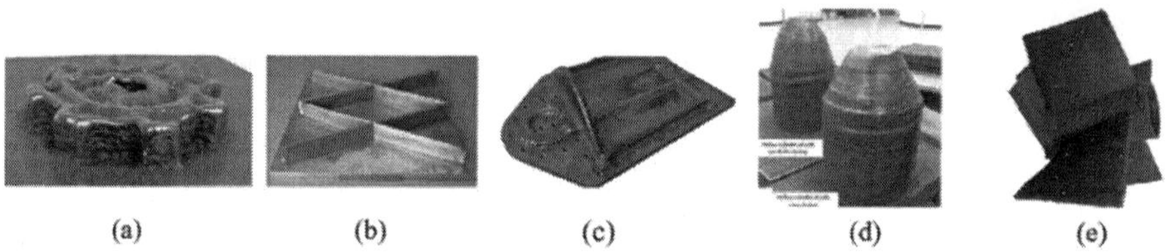

FIGURE 2.3 Typical WAAM products: (a) ER70S-6 steel gear, (b) aluminium alloy panels, (c) Airbus A320 aerial support made of Ti-6Al-4V, (d) cone-shaped components made of ER-70S-6 steel and ER316L, and (e) propeller [25].

FIGURE 2.4 Typical thin-walled steel components fabricated by WAAM [25].

adopted to manufacture diverse structures and allows for expanding the design of products, saving material utilisation and production costs.

2.3 A CASE STUDY ON WAAM OF LOW-CARBON STEEL PARTS

2.3.1 Materials and Sample Preparation

Herein, a thin-walled component was produced by GMAW-WAAM. 1.2-mm-in-diameter ER70S-6 mild steel wire was utilised as the feedstock. The GMAW-AM system consists of a 6-axis robot (1), which performs the deposition paths for the welding torch (4) (Figure 2.5a). The robot controller (2) provides control signals to robots. The welding source (3) controls the welding parameters.

The thin wall has been built on a low-carbon steel substrate utilising the following parameters: travel speed = 0.3 m/min, voltage = 18 V, and current = 90 A. The contamination and oxide layer on the substrate's surface were removed. The wall was built using a zigzag deposition path with an idle time of 45 seconds between two successive layers (Figure 2.5b). The shielding gas is CO_2, with a flow rate of 14–15 L/min.

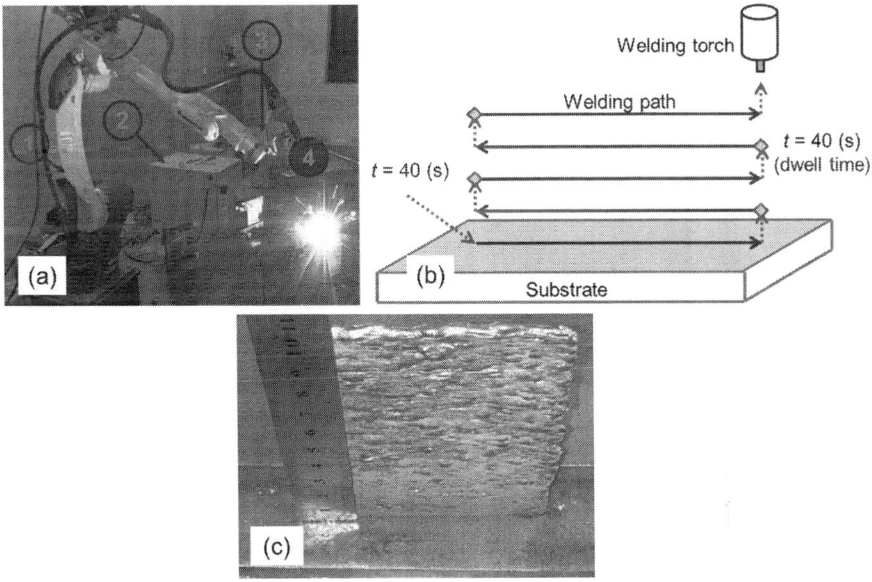

FIGURE 2.5 (a) The GMAW-WAAM system, (b) the deposition paths, and (c) the as-built thin-walled sample.

2.3.2 Characterisation Methods

To observe the microstructures and microhardness of the wall (Figure 2.5c), three samples, from the top to the bottom zones, were cut. They were polished and etched with 5% Nital solution. The microstructures were analysed using optical microscopes. The microhardness (HV) was tested utilising a Vickers hardness machine. Each test was performed with 49.05 N in load and 10 seconds in dwell time.

Three tensile samples were extracted in each direction – the deposition (vertical) direction Z and the welding (horizontal) direction X (Figure 2.6a). The specimens had been designed according to ASTM A370 standard (Figure 2.6b). The bond strength between the deposited material and the substrate was also evaluated. For this task, three specimens (b-Tsi) were prepared (Figure 2.6c). The tensile test was performed with a rate of 1 mm/min on the Instron 3369 machine.

2.3.3 Microstructure Characterisation

The typical microstructures of straight, thin-walled components made of low-carbon steels (ER70-S6) are shown in Figure 2.7. The microstructure varies along the bottom-to-top direction of the wall. The top zone reveals lamellar morphology (Figure 2.7a). In contrast, granular grains of ferrite with a small quantity of perlite at the ferrite grains' boundary appear in the centre (Figure 2.7b). The bottom zone shows a mixture of equiaxed and lamellar structures (Figure 2.7c). The temperature transmission along the top-to-bottom direction of the part results in the austenite dendrites transformed into Widmanstädter ferrite (α_w) structures.

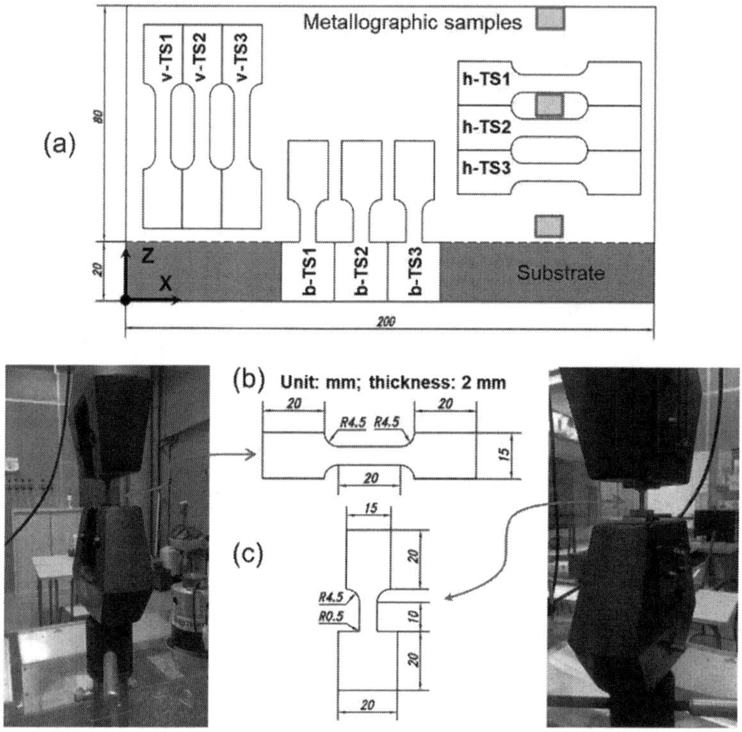

FIGURE 2.6 (a) Locations for extracting tensile samples, (b), and (c) the samples' dimensions.

The microstructures of the centre zone are also distinguished into two types: equiaxed structures with dense distributions in the non-overlapping region (Figure 2.7b-1) and granular structures in the overlapping region with larger sizes (Figure 2.7b-2). Because of the heat effect of the currently deposited layer, the previously deposited layer is reheated and remelted. Hence, solid-state phase transformation occurs, such as grain growth, recrystallisation, phase transitions, annealing, and tempering in the overlapping layers, resulting in coarser grain size when compared with the non-overlapped zones. The microstructures in the bottom zone are the result of the higher cooling rate due to it being close to the substrate at room temperature, while the centre zone lies on hot layers [45].

2.3.4 Mechanical Properties

The measured microhardness and tensile strengths are comparable to corresponding wrought steels and suitable for practical applications [46]. The top zone reveals the highest average value (205 ± 5 HV), followed by the bottom (195 ± 7 HV) and the middle zone (176 ± 2 HV).

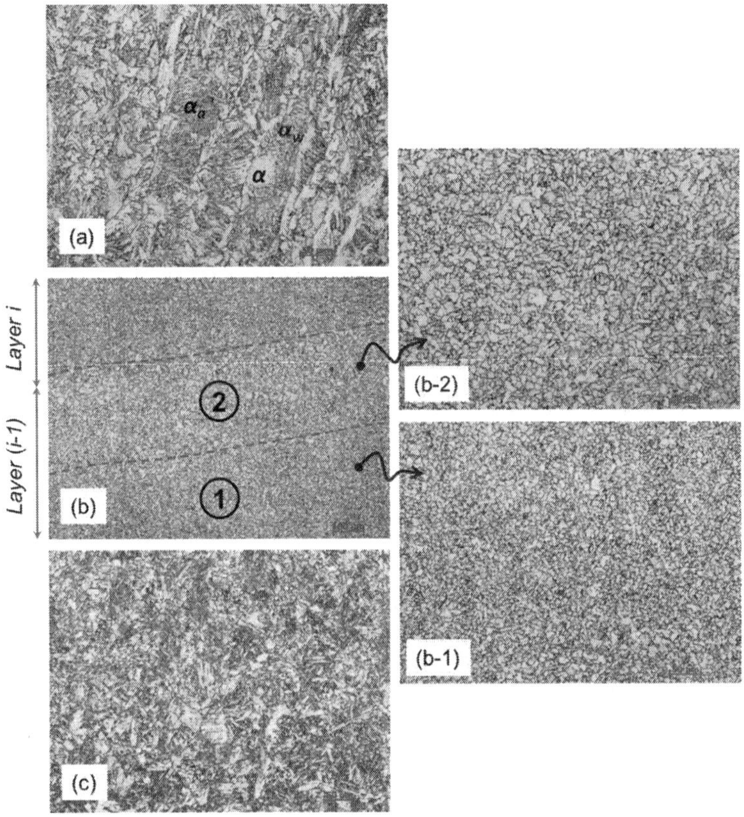

FIGURE 2.7 Microstructures in (a) top, (b) middle, and (c) bottom zones of the thin wall.

Tensile tests are carried out in three zones (Figure 2.6a) with longitudinal, horizontal, and longitudinal directions adjacent to the substrate. Tensile strengths (YS and UTS) of the samples (v-TSi) in the deposition (vertical) direction are higher than those of the horizontal samples (h-TSi) (Figure 2.8). The UTS and YS values of vertical samples are 479 ± 7 and 362 ± 8 MPa. On the other hand, the tensile properties of the h-TSi samples are UTS = 429 ± 8 MPa and YS = 320 ± 6 MPa. The difference in tensile strength and elongation between the vertical and horizontal directions is caused by microstructure variation.

The b-TSi samples all fractured in the middle zone instead of in the region adjacent to the substrate, expressing a high-bonding strength of the deposited metal with the base metal, and that meets working requirements in practice. The average values of UTS and YS of b-TSi samples are 468 ± 8 MPa and 355 ± 7 MPa, which are close to those of the longitudinal samples v-TSi.

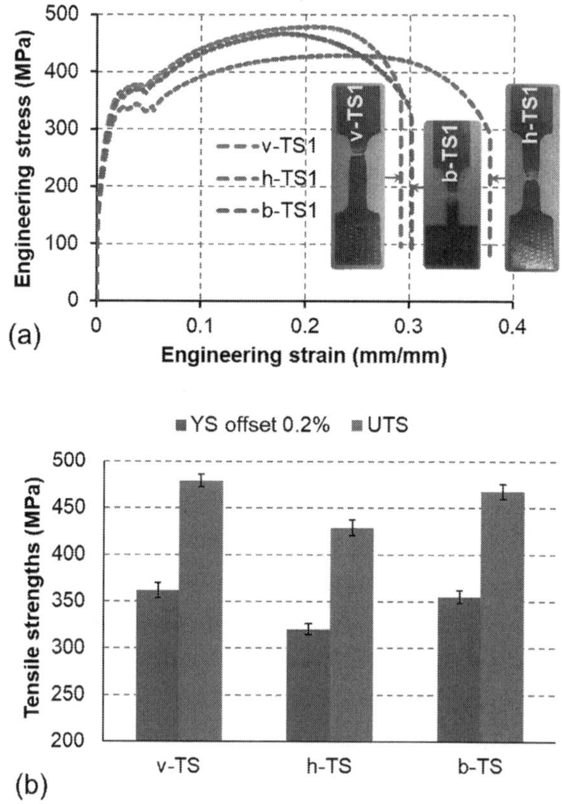

FIGURE 2.8 (a) Strain-stress curve and (b) average values of tensile properties.

2.4 CONCLUSION

This chapter first presents an overview of the WAAM technology, which currently has increasing attention for large-sized components. This technology's definition, classification, advantages, and disadvantages have been presented. The chapter also presents the typical defects of as-built WAAM components and the available solutions to mitigate existing defects. Secondly, a case study on the WAAM process of low-carbon steel components was discussed. The results revealed that thin walls feature variations in microstructure and can be distinguished in three regions. This phenomenon leads to heterogeneity in mechanical properties. However, the mechanical characteristics of as-built low-carbon steel walls are close to that of corresponding wrought materials and meet the requirements of practical applications. In summary, the chapter provides a basic understanding of the WAAM technology with a case study on the manufacture of low-carbon steel parts, which contributes to valuable knowledge for industrial and academic communities.

NOTE

1 http://additive-manufacturing-report.com

REFERENCES

1. Vayre B, Vignat F, Villeneuve F (2013) Identification on some design key parameters for additive manufacturing: Application on electron beam melting. *Procedia CIRP* 7: 264–269. https://doi.org/10.1016/j.procir.2013.05.045
2. Guo N, Leu M (2013) Additive manufacturing: Technology, applications and research needs. *Frontiers of Mechanical Engineering* 8: 215–243. https://doi.org/10.1007/s11465-013-0248-8
3. ASTM (2010) F2792 – 10e1 Standard Terminology for Additive Manufacturing Technologies. https://doi.org/10.1520/F2792-10E01
4. Le VT, Paris H (2018) A life cycle assessment-based approach for evaluating the influence of total build height and batch size on the environmental performance of electron beam melting. *The International Journal of Advanced Manufacturing Technology* 98: 275–288. https://doi.org/10.1007/s00170-018-2264-7
5. Le VT, Paris H, Mandil G (2017) Environmental impact assessment of an innovative strategy based on an additive and subtractive manufacturing combination. *Journal of Cleaner Production* 164: 508–523. https://doi.org/10.1016/j.jclepro.2017.06.204
6. Huang R, Riddle M, Graziano D, et al. (2016) Energy and emissions saving potential of additive manufacturing: The case of lightweight aircraft components. *Journal of Cleaner Production* 135: 1559–1570. https://doi.org/10.1016/j.jclepro.2015.04.109
7. Le Bourhis F, Kerbrat O, Hascoet J-Y, Mognol P (2013) Sustainable manufacturing: Evaluation and modeling of environmental impacts in additive manufacturing. *The International Journal of Advanced Manufacturing Technology* 69: 1927–1939. https://doi.org/10.1007/s00170-013-5151-2
8. Paris H, Mokhtarian H, Coatanéa E, et al. (2016) Comparative environmental impacts of additive and subtractive manufacturing technologies. *CIRP Annals – Manufacturing Technology* 65: 29–32. https://doi.org/10.1016/j.cirp.2016.04.036
9. Suárez A, Aldalur E, Veiga F, et al. (2021) Wire arc additive manufacturing of an aeronautic fitting with different metal alloys: From the design to the part. *Journal of Manufacturing Processes* 64: 188–197. https://doi.org/10.1016/j.jmapro.2021.01.012
10. Omiyale BO, Olugbade TO, Abioye TE, Farayibi PK (2022) Wire arc additive manufacturing of aluminium alloys for aerospace and automotive applications: A review. *Materials Science and Technology* 38: 391–408. https://doi.org/10.1080/02670836.2022.2045549
11. Priarone PC, Ingarao G (2017) Towards criteria for sustainable process selection: On the modelling of pure subtractive versus additive/subtractive integrated manufacturing approaches. *Journal of Cleaner Production* 144: 57–68. https://doi.org/10.1016/j.jclepro.2016.12.165
12. Serres N, Tidu D, Sankare S, Hlawka F (2011) Environmental comparison of MESO-CLAD process and conventional machining implementing life cycle assessment. *Journal of Cleaner Production* 19: 1117–1124. https://doi.org/10.1016/j.jclepro.2010.12.010
13. Le VT, Paris H, Mandil G (2018) The development of a strategy for direct part reuse using additive and subtractive manufacturing technologies. *Additive Manufacturing* 22: 687–699. https://doi.org/10.1016/j.addma.2018.06.026
14. Herzog D, Seyda V, Wycisk E, Emmelmann C (2016) Additive manufacturing of metals. *Acta Materialia* 117: 371–392. https://doi.org/10.1016/j.actamat.2016.07.019

15. Dutta B, Froes FH (2015) *The additive manufacturing (AM) of titanium alloys*. Elsevier Inc.
16. Frazier WE (2014) Metal additive manufacturing: A review. *Journal of Materials Engineering and Performance* 23: 1917–1928. https://doi.org/10.1007/s11665-014-0958-z
17. Oliveira JP, LaLonde AD, Ma J (2020) Processing parameters in laser powder bed fusion metal additive manufacturing. *Materials & Design* 193: 108762. https://doi.org/10.1016/j.matdes.2020.108762
18. Mandil G, Le VT, Paris H, Suard M (2016) Building new entities from existing titanium part by electron beam melting: Microstructures and mechanical properties. *The International Journal of Advanced Manufacturing Technology* 85: 1835–1846. https://doi.org/10.1007/s00170-015-8049-3
19. Körner C (2016) Additive manufacturing of metallic components by selective electron beam melting – a review. *International Materials Reviews* 6608: 1–17. https://doi.org/10.1080/09506608.2016.1176289
20. Williams SW, Martina F, Addison AC, et al. (2016) Wire + arc additive manufacturing. *Materials Science and Technology* 32: 641–647. https://doi.org/10.1179/1743284715Y.0000000073
21. Li Y, Su C, Zhu J (2022) Comprehensive review of wire arc additive manufacturing: Hardware system, physical process, monitoring, property characterization, application and future prospects. *Results in Engineering* 13: 100330. https://doi.org/10.1016/j.rineng.2021.100330
22. Cunningham CR, Flynn JM, Shokrani A, et al. (2018) Invited review article: Strategies and processes for high quality wire arc additive manufacturing. *Additive Manufacturing* 22: 672–686. https://doi.org/10.1016/j.addma.2018.06.020
23. Le VT, Mai DS, Paris H (2021) Influences of the compressed dry air-based active cooling on external and internal qualities of wire-arc additive manufactured thin-walled SS308L components. *Journal of Manufacturing Processes* 62: 18–27. https://doi.org/10.1016/j.jmapro.2020.11.046
24. Le VT, Mai DS, Bui MC, et al. (2022) Influences of the process parameter and thermal cycles on the quality of 308L stainless steel walls produced by additive manufacturing utilizing an arc welding source. *Welding in the World* 66: 1565–1580. https://doi.org/10.1007/s40194-022-01330-4
25. Jafari D, Vaneker THJ, Gibson I (2021) Wire and arc additive manufacturing: Opportunities and challenges to control the quality and accuracy of manufactured parts. *Materials & Design* 202: 109471. https://doi.org/10.1016/j.matdes.2021.109471
26. Wu B, Pan Z, Ding D, et al. (2018) A review of the wire arc additive manufacturing of metals: Properties, defects and quality improvement. *Journal of Manufacturing Processes* 35: 127–139. https://doi.org/10.1016/j.jmapro.2018.08.001
27. Jin W, Zhang C, Jin S, et al. (2020) Wire arc additive manufacturing of stainless steels: A review. *Applied Sciences (Switzerland)* 10: https://doi.org/10.3390/app10051563
28. Le VT, Mai DS, Doan TK, Paris H (2021) Wire and arc additive manufacturing of 308L stainless steel components: Optimization of processing parameters and material properties. *Engineering Science and Technology, an International Journal* 24: 1015–1026. https://doi.org/10.1016/j.jestch.2021.01.009
29. Kumar N, Bhavsar H, Mahesh PVS, et al. (2022) Wire arc additive manufacturing – A revolutionary method in additive manufacturing. *Materials Chemistry and Physics* 285: 126144. https://doi.org/10.1016/j.matchemphys.2022.126144
30. Raut LP, Taiwade RV (2021) Wire arc additive manufacturing: A comprehensive review and research directions. *Journal of Materials Engineering and Performance* 30: 4768–4791. https://doi.org/10.1007/s11665-021-05871-5

31. da Silva LJ, Souza DM, de Araújo DB, et al. (2020) Concept and validation of an active cooling technique to mitigate heat accumulation in WAAM. *International Journal of Advanced Manufacturing Technology* 107: 2513–2523. https://doi.org/10.1007/s00170-020-05201-4
32. Derekar KS (2018) A review of wire arc additive manufacturing and advances in wire arc additive manufacturing of aluminium. *Materials Science and Technology* 34: 895–916. https://doi.org/10.1080/02670836.2018.1455012
33. Xia C, Pan Z, Polden J, et al. (2020) A review on wire arc additive manufacturing: Monitoring, control and a framework of automated system. *Journal of Manufacturing Systems* 57: 31–45. https://doi.org/10.1016/j.jmsy.2020.08.008
34. Liu J, Xu Y, Ge Y, et al. (2020) Wire and arc additive manufacturing of metal components: A review of recent research developments. *International Journal of Advanced Manufacturing Technology* 111: 149–198. https://doi.org/10.1007/s00170-020-05966-8
35. Montevecchi F, Venturini G, Grossi N, et al. (2018) Idle time selection for wire-arc additive manufacturing: A finite element-based technique. *Additive Manufacturing* 21: 479–486. https://doi.org/10.1016/j.addma.2018.01.007
36. Takagi H, Sasahara H, Abe T, et al. (2018) Material-property evaluation of magnesium alloys fabricated using wire-and-arc-based additive manufacturing. *Additive Manufacturing* 24: 498–507. https://doi.org/10.1016/j.addma.2018.10.026
37. Yi H-J, Kim J-W, Kim Y-L, Shin S (2020) Effects of cooling rate on the microstructure and tensile properties of wire-arc additive manufactured Ti–6Al–4V alloy. *Metals and Materials International*. https://doi.org/10.1007/s12540-019-00563-1
38. Hackenhaar W, Montevecchi F, Scippa A, Campatelli G (2019) Air-cooling influence on wire arc additive manufactured surfaces. *Key Engineering Materials* 813 KEM: 241–247. https://doi.org/10.4028/www.scientific.net/KEM.813.241
39. Wu B, Pan Z, Ding D, et al. (2018) The effects of forced interpass cooling on the material properties of wire arc additively manufactured Ti6Al4V alloy. *Journal of Materials Processing Technology* 258: 97–105. https://doi.org/10.1016/j.jmatprotec.2018.03.024
40. Reisgen U, Sharma R, Mann S, Oster L (2020) Increasing the manufacturing efficiency of WAAM by advanced cooling strategies. *Welding in the World* 64: 1409–1416. https://doi.org/10.1007/s40194-020-00930-2
41. Wu B, Pan Z, Chen G, et al. (2019) Mitigation of thermal distortion in wire arc additively manufactured Ti6Al4V part using active interpass cooling. *Science and Technology of Welding and Joining*: 1–11. https://doi.org/10.1080/13621718.2019.1580439
42. Ding D, Wu B, Pan Z, et al. (2020) Wire arc additive manufacturing of Ti6AL4V using active interpass cooling. *Materials and Manufacturing Processes* 35: 845–851. https://doi.org/10.1080/10426914.2020.1732414
43. Hackenhaar W, Mazzaferro JAE, Montevecchi F, Campatelli G (2020) An experimental-numerical study of active cooling in wire arc additive manufacturing. *Journal of Manufacturing Processes* 52: 58–65. https://doi.org/10.1016/j.jmapro.2020.01.051
44. Le VT, Mai DS, Hoang QH (2020) Effects of cooling conditions on the shape, microstructures, and material properties of SS308L thin walls built by wire arc additive manufacturing. *Materials Letters* 280: 128580. https://doi.org/10.1016/j.matlet.2020.128580
45. Liberini M, Astarita A, Campatelli G, et al. (2017) Selection of optimal process parameters for wire arc additive manufacturing. *Procedia CIRP* 62: 470–474. https://doi.org/10.1016/j.procir.2016.06.124
46. ASTM. ASTM A36 Steel, plate. http://www.matweb.com/search/DataSheet.aspx?MatGUID=afc003f4fb40465fa3df05129f0e88e6&ckck=1

3 Process Planning and Parameters Selection in Wire Arc Additive Manufacturing

Bunty Tomar and S. Shiva
Indian Institute of Technology Jammu, India

3.1 INTRODUCTION

Wire and arc-based additive manufacturing (WAAM) has the capability of manufacturing bulk and big components by the high rate of material deposition, effective raw material utilisation, lesser infrastructure cost, and consequent environmental friendliness. These capabilities of WAAM have attracted the industrial manufacturing sector more and more in recent years [1]. The patent of WAAM was filed by Ralph in 1925 when he used an electric arc as the source of fusion energy and metallic wire as the raw material for creating metallic decorations [2]. In this technology, the wire feeding mechanism and the arc welding principle are integrated. By utilising the provided 3D computer-aided design (CAD) model data, the wired feed material is melted, employing the heat of the created arc, and the molten metal is principally deposited on a specified substrate in a layer-upon-layer manner. A robot-assisted arm and gantry system machine uses this continuous layer-by-layer deposition to create the entire 3D metal structure [3].

Gas metal arc welding (GMAW) [4], gas tungsten arc welding (GTAW) [5], and plasma arc welding (PAW) [6] are three main categories of WAAM based upon the type of energy source used during the process. The primary determining element in choosing the precise welding source for a given WAAM application is the user requirement. If the application demands a greater deposition rate, GMAW-based WAAM could be utilised instead of GTAW-based WAAM. For tracks with less distortion, PAW-WAAM should be preferred over GMAW-WAAM. The choice of welding methods in the WAAM technique affects the amount of time required, the rate of deposition, and the necessary processing parameters for a particular alloy [7]. Reference [7] presents a summarised detail of all these variants of the WAAM technique. Cold metal transfer-based wire arc additive manufacturing (CMT-WAAM) [8] is another variant of WAAM based on GMAW-WAAM, and its unique characteristics of providing controlled and lesser heat input and lesser spatter excels it over other conventional WAAM techniques [9].

DOI: 10.1201/9781003363415-3

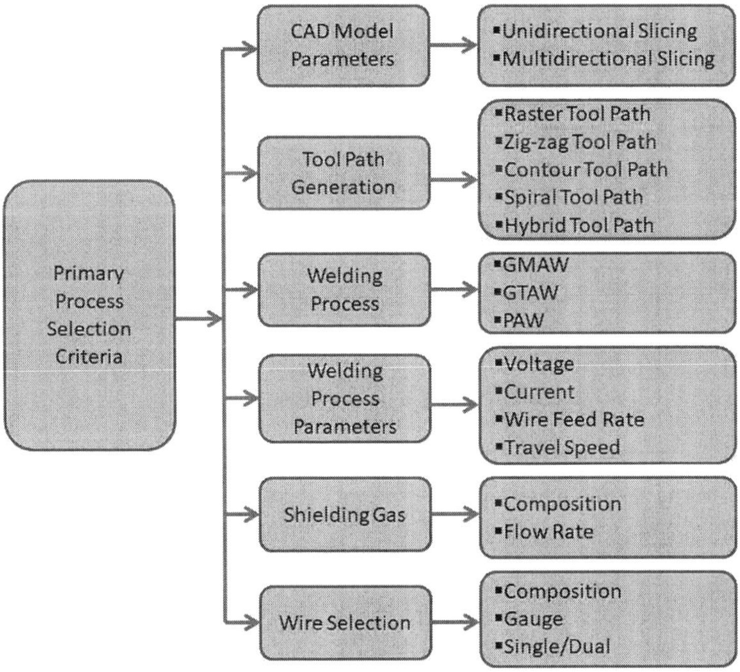

FIGURE 3.1 Main process parameters employed in WAAM [7]. (Publishing permitted by Elsevier.)

The WAAM approach has the capacity to create application-based components with near-net-shape and surface roughness that is tolerably acceptable. It is necessary to machine end-use components using the proper machining techniques. A number of procedure selection factors and computer-aided process planning must be taken into consideration in order to determine WAAM's capacity to manufacture almost flawless components made of certain materials. The CAD parameters, tool path generation, joining method, joining parameters, shielding gas, and feedstock selection are the main selection factors for processes. This chapter begins with a summary of process planning parameters employed in WAAM and discusses their selection techniques. Further, the critical factors for the employment of these process parameters in the WAAM technique and their effects on the microstructure and mechanical properties of fabricated components are detailed in the chapter. Figure 3.1 presents the primary process parameters in WAAM [7].

3.2 CAD MODEL PARAMETERS

For the fabrication of all additive manufacturing (AM) processes, CAD model design, 2D slicing of generated 3D CAD design, and tool path development are required planning procedures. The component's geometry should ideally be reverse-engineered and saved in the common ".stl" file format so that a 3D CAD model may be created and used to slice the component [10]. Because of its simplicity,

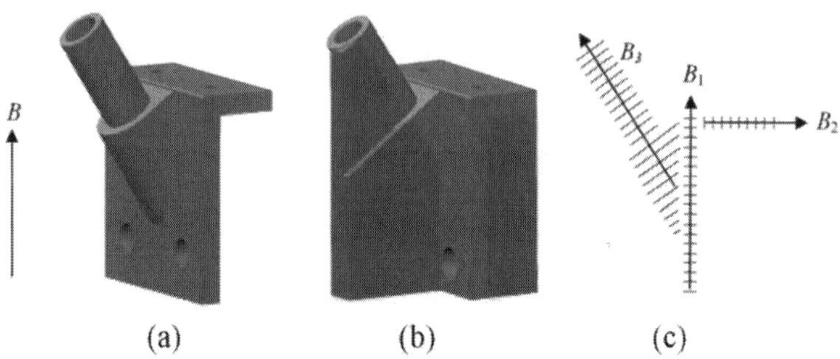

FIGURE 3.2 (a) Component with build direction (BD) B, (b) showing the support requirement (coloured), (c) B1, B2, and B3 BDs for MDS [11]. (Publishing permitted by Elsevier.)

a unidirectional slicing approach is now employed in AM machines. The usage of the unidirectional slicing approach is somewhat constrained for large designs since it necessitates several support systems for intricate geometries, as well as overhanging components. The multidirectional slicing (MDS) approach is utilised in newly established AM machinery to reduce the implementation of support structures in complicated structures and overhanging geometries. By slicing in MDS, it is feasible to deposit material at the underside of an overhang because a multi-axis robotic setup may deposit the feedstock over the base surface in the desired orientation [11, 12]. Because of the complexity of its algorithms and the requirement for multidirectional depositing nozzles, MDS necessitates a full robotic configuration rig [13]. A display of the MDS method is presented in Figure 3.2.

3.3 TOOL PATH GENERATION

Another crucial stage in the WAAM process planning is the detailed tool route strategy. In WAAM, the creation of the tool path is essential because it directs the deposition nozzle's movement to deposit in sliced 2D layers that represent the component's cross-sections. The process of offsetting is crucial to the creation of tool paths. The most popular routes for scanning procedures in WAAM are raster and contour patterns. Offsetting is carried out parallel to a direction when using a raster pattern but parallel to the structure's edge when using a contour pattern. The remaining route patterns are either iterations of these scanning algorithms or combinations of them [10]. Figure 3.3 shows various created tool paths in this regard. The raster technique based on the planer beam's projection in one direction is seen in Figure 3.3a. With this technique, a large number of scan lines with tiny widths that fill 2D space are projected [14]. Raster path patterns have a core component that is constructed in a certain direction, which necessitates many rotations of the deposition head and leads to subpar construction. The one continuous pass linking two parallel scanning lines distinguishes the zigzag filling route from raster scanning, as seen in Figure 3.3b. Due to edge discretisation errors and deflection in alternating directions, both of these techniques exhibit low outer boundary accuracy [15, 16]. As seen in Figure 3.3c, the contour path follows the deposition path from the border to the interior.

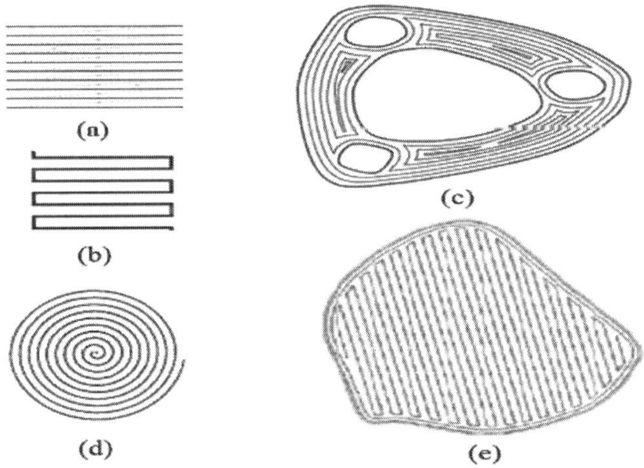

FIGURE 3.3 Various tool path geometries used (a) raster, (b) zigzag, (c) contour, (d) spiral, and I hybrid [10]. (Publishing permitted by Springer.)

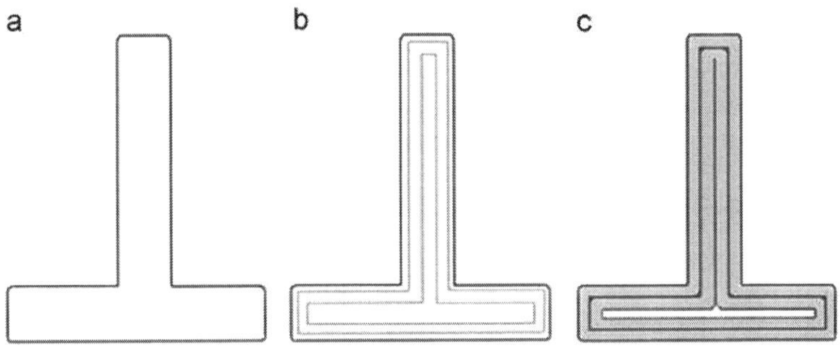

FIGURE 3.4 (a) Cross-sectional view of the thin wall, (b) contour tool path, and (c) material deposited with unfiled zone [10]. (Publishing permitted by Elsevier.)

For manufacturing thin-walled metallic components, contour path strategies are typically favoured over raster and zigzag strategies since they don't necessitate turning the deposition head frequently. Moreover, by continuously altering the route direction along the border, contour planning-based manufactured components get around the issues of distortion and mechanical anisotropy [17]. The contour path pattern causes difficulty since it starts filling from the outside in and does not guarantee the entire filling of a specified 2D shape, as seen in Figure 3.4. This results in empty parts inside the deposited layers [10]. Figure 3.4 displays the cross-section corresponding to a thin-walled geometry (a). By displacing the border, contour routes (shown in Figure 3.4(b) as green lines) are created to fill this area by AM. When "d" is the hatch spacing, the provided offset to the "n"th layer is located at the distance of (n-0.5d) from the edge to its interior (step-over distance). According to Figure 3.5, the step-over distance is the separation among the centres of the next and preceding deposition tracks. Weld beads will inevitably overlap in order to provide a smooth surface

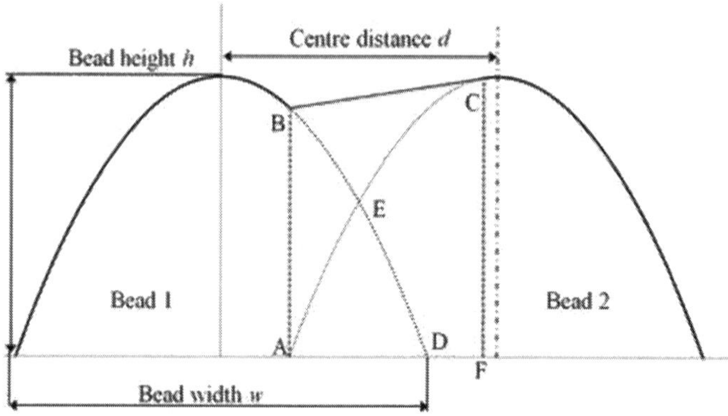

FIGURE 3.5 Display of hatch spacing (step-over distance d) [18]. (Publishing permitted by Elsevier.)

in WAAM. According to a number of weld bead overlapping models, the ideal hatch spacing in WAAM depends on the bead width 'w' in a way that d = 0.667 w or d = 0.738 w [18]. The narrow empty space remained by contour path scanning at the point of deposition (the white area in Figure 3.4c) is discovered because there isn't enough room for the subsequent contour path to be offset. The amount of empty space is solely dependent on the step-over distance d, which varies for various AM techniques. In WAAM, d is dependent on the feed diameter, deposition head travel speed, and wire feed rate. Depending on the application, its usual value falls between 2 and 12 mm. Such empty areas run the risk of causing heavily loaded constructions to fail early. Decreasing hatch spacing might be one way to prevent these empty spaces; however, in WAAM, this would slow down the rate of deposition. Revisiting the gaps by overlapping the deposition path is another explanation that is conceivable; however, doing so may result in structures with thicker walls. A different approach called medial axis transformation (MAT) is suggested to overcome this problem [19].

Due to its advantages over the zigzag tool path, the spiral path planning (given in Figure 1.3d) is appropriate in several unusual geometrical forms. This technique may also be used in AM; however, it was initially marketed in machining procedures like pocket milling [20]. The contour route and zigzag path planning are combined to create a hybrid tool path, which addresses the drawbacks of the previously stated tool path schemes (shown in Figure 3.3e). In this design, the inner core is filled using a zigzag tool path, while the outer core is scanned using spiral path patterns. The hybrid tool path is the most advantageous type of tool path planning approach in WAAM because it integrates a number of positive qualities that the other tool path strategies are unable to perform [17, 21].

3.4 WELDING PROCESS

GMAW, GTAW, and PAW are the welding techniques that are employed in WAAM. The end-use application, as well as ease of process implementation, determine which welding technique is chosen. For example, GMAW should be chosen over GTAW

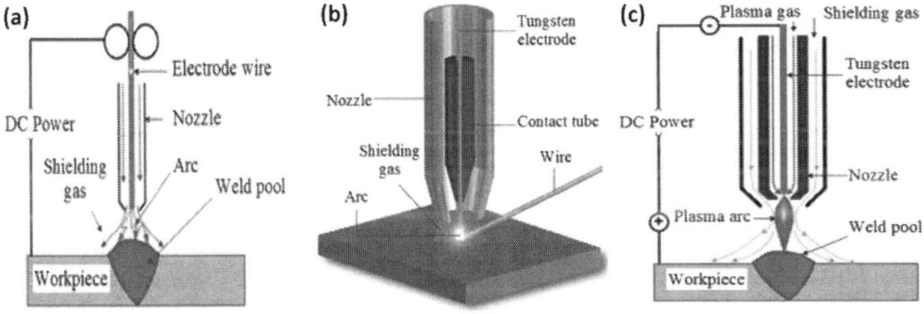

FIGURE 3.6 Schematic representation of (a) GMAW, (B) GTAW, and (c) PAW-based techniques [13]. (Open access.)

and PAW for manufacturing if material deposition rate is a priority, despite the fact that it does not offer the greatest fabrication quality and process stability. Because GMAW doesn't require external wire feeding, it is more straightforward to deploy than GTAW and PAW. PAW's high energy density electrical arc produces the best superiority welds with exact bead form and size and little distortion [22]. Because of the high energy density of the PAW electric arc, better tensile characteristics are achieved by employing PAW-based WAAM fabrication of steels than GMAW. PAW technique results in the least amount of bead distortion. According to this, the necessity for end-use qualities may also influence the choice of welding method [23]. GMAW is chosen over other techniques for the wire being supplied coaxially to the welding flame, in addition to the high deposition rate. As per the ASTM F3187, feeding wire coaxially provides a simpler tool path, and GMAW stands out as a superior alternative to GTAW and PAW techniques because GMAW doesn't need a particular set-up for reorienting the feedstock as well as deposition direction [24, 25]. Yet, by maintaining the wire feed angle of 60°, geometrically uniform fabrication is possible in all travel directions [26]. Figure 3.6 presents a schematic representation of the basic principle and working of GMAW, GTAW, and PAW-based processes.

3.5 WELDING PROCESS PARAMETERS

Many welding process factors, such as voltage, circuit current, wire feed speed (WFS), and travel rate (TR), should be managed and regulated in order to produce stable and defect-free structures using wire and arc AM. As such factors collectively affect the heat input in the deposit, they will directly affect the weld properties and quality of the deposited structure [27]. The macrostructure and weld geometry will worsen as a result of an excessive heat input during deposition, which will cause previously deposited layers to re-melt to a greater extent. A homogeneous deposition surface may be achieved with a reduced equivalent heat input without the weld pool overflowing and collapsing [28]. By lowering the drop temperature and the degree of gas solubility in the molten pool, lesser heat input is also helpful in lowering the total porosity in the fabricated component [29]. A study investigating the effect of WFS and TR on heat input showed that a rise in the ratio of WFS to TR is responsible for an increase in heat input [30]. In WAAM, the process aim is the reduction in the amount of heat input while increasing the deposition rate. WFS is exactly dependent on given heat input, which raises even more concerns about this scenario. According to research on PAW-based Ti64 WAAM, steady bead deposition is best successful by maintaining a WFS/TR ratio of 30 [24].

The melt pool's size is significantly influenced by the heat input in the WAAM. The width-to-height ratio leads towards lesser with lower heat inputs because the molten pool doesn't get enough time for expansion prior to solidifying. The viscosity of the metal is reduced by increasing the heat input, making it easier to extend the melt pool. Yet, by incorporating the cooling power into the heat input, it is possible to reduce the difference in heat dissipation among two successive layers and produce a constant bead shape [31]. Synergic welding programmes assist in providing a stable welding process in welding specific alloys when used in single-pass welding or for substrates at atmospheric temperature. During a WAAM build, these synergistic programmes could no longer be suitable for heat build-up.

Another important factor in arc welding procedures is arc current, which has a big impact on bead roughness. It also regulates the weld shape, depth of fusion, and electrode burn-off rate [32]. Stronger welded joints are produced by increasing the weld penetration depth, which is also caused by a greater current rate [33]. The fusion zone phase of the deposit is determined by the potential difference between the melt pool and the feedstock. The penetrating depth is also influenced by arc voltage since welds with a significantly greater arc voltage tend to be broader, flatter, and less deeply penetrated. The best penetration depth is achieved by using an appropriate ideal arc voltage [33]. At a specific arc voltage, the penetration is greatest. Once the arc voltage is increased past this ideal point, penetration will begin to decline. Figure 3.7 provides a schematic representation of how to regulate the welding current

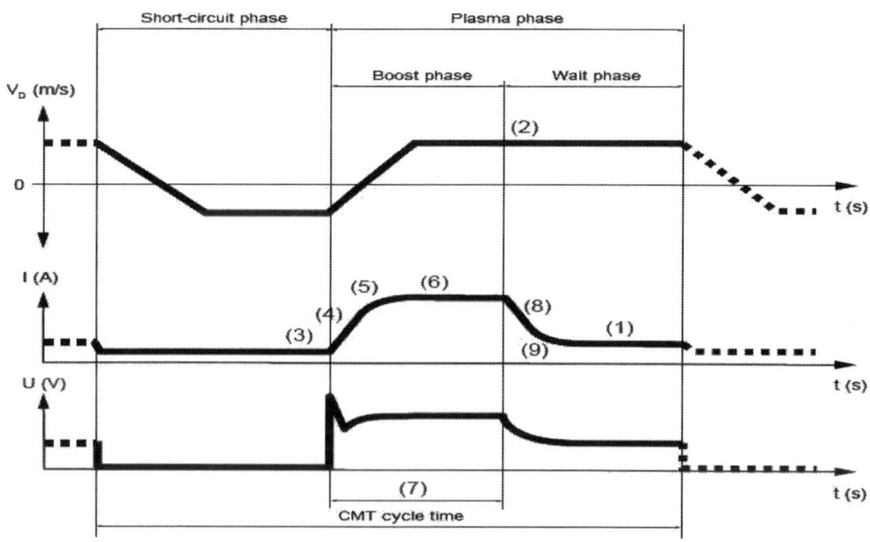

FIGURE 3.7 Current and voltage waveforms for the CMT-WAAM cycle [34]. (Open access.)

and voltage to regulate the heat input and the rate of material deposition in cold metal transfer-based WAAM [34].

3.6 SHIELDING GAS

A sufficient amount of shielding gas must be used to shield the area around the hot, molten material to prevent contamination in the weld pool during WAAM deposition. It averts the probability of certain flaws like weld cracking, porosity, and inclusions and also stops harmful oxides and nitrides from forming [35]. Poor penetration might be caused by a greater shielding gas flow rate. Due to trapped air gases caused by high flow rate turbulence, the final component may be porous. These flaws are frequently discovered in the vicinity of the welding set-up when ambient gases, including oxygen (O_2), nitrogen (N_2), and water vapour (H_2O), are present. In general situations, shielding gas is provided by a welding torch, although specific preparations for the shielding gas supply are established for the deposition of very contaminant-prone materials like Ti64 and maraging steel [36]. Figure 3.8 depicts a wall of maraging steel made by WAAM that has been oxidation-degraded for torch-only shielding. As shown in Figure 3.9, a tent filled with argon was utilised in the same investigation instead of torch shielding and performed very well in means of material deposition, surface roughness, and manufacturing efficiency [37]. With WAAM built of Ti64

FIGURE 3.8 Oxide accumulation on WAAM sample of maraging steel for torch-only shielding [37]. (Publishing permitted by Elsevier.)

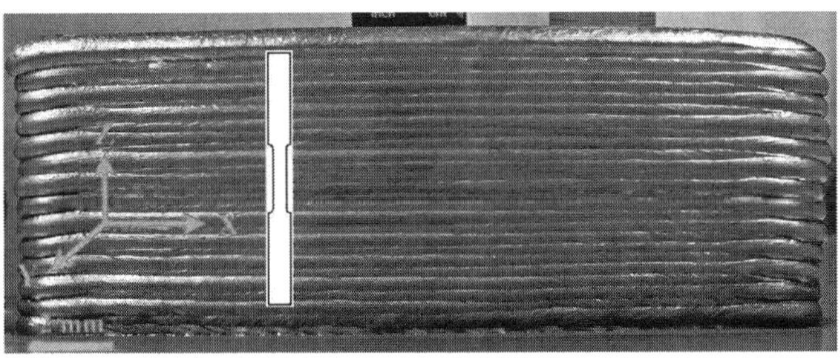

FIGURE 3.9 Argon-filled, tent-shielded WAAM fabricated sample of maraging steel [37]. (Publishing permitted by Elsevier.)

material, a local shielding machine that provides the laminar flow to the shielding gas is created to offer superior shielding conditions than traditional shielding [38].

Another crucial consideration in the procedure design of a wire and arc AM set-up for minimising arc wandering for exceedingly sensitive materials, given aluminium and Ti alloys, is the composition of the shield gas. Lesser accessibility of the oxides inside the melt pool causes arc wandering in the case of aluminium and its alloys [39]. The arc travels towards the solid surface, looking for additional oxides at times of inadequate oxide availability on the molten pool surface, creating a distinctive imprint over the substrate plate known as a sputtering zone, as seen in Figure 3.10 [40]. There are various advantages to using an oxygen-doped shielding gas, such as faster travel, better weld penetration, and reduction in molten metal surface tension [41]. For O_2 concentrations of 0.3, 3, 200, and 2,000 ppm in Ar-based shielding gas in WAAM of a thin aluminium layer, the arc cathodic emission behaviour was examined. Figure 3.11 displays the comparison frames of arc behaviour for various shielding gas O_2 contents [42]. The findings indicated that the narrowest sputtering zone with minimum arc wandering is formed by argon doped with 200 ppm of oxygen gas in a shielding gas [40, 42]. Shielding gas with the composition of Ar and He content was used to fabricate Ti64 samples with refined previous grains because it aided in accelerating cooling in the melt pool surface [43].

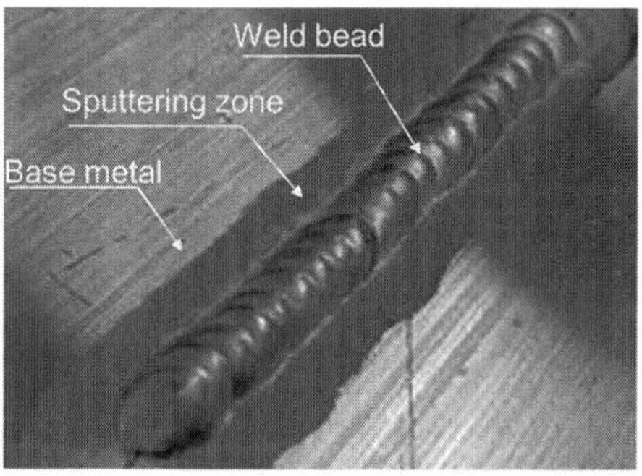

FIGURE 3.10 Sputtering zone in the WAAM [40]. (Publishing permitted by Elsevier.)

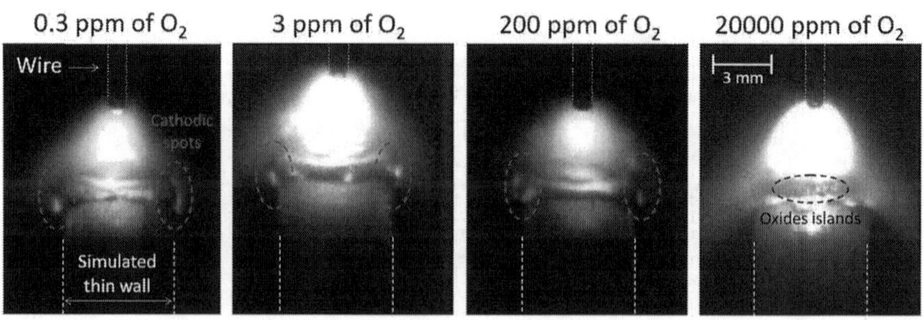

FIGURE 3.11 Arc behaviour for WAAM of Al alloy with varying oxygen content in shielding gas [42]. (Publishing permitted by Elsevier.)

3.7 FEEDSTOCK

The choice of feedstock during the WAAM method planning is crucial in developing different performance characteristics. The rates of deposition heat transfer and general weld quality are all influenced by the feedstock dia. and the quantity of feedstocks fed in the produced arc [44]. It is feasible to use more than one wire in wire arc AM of in situ alloying. It was discovered that using dual wire for wire arc AM of Al-Cu-Sn alloy improved surface quality, microstructure, and strength. The superheating degree of melt produced by the dual-wire process was significantly lower than the single-wire molten pool when processing conditions were the same because the heat generated in dual-wire WAAM was practically half that of the single-wire WAAM. An arc that was more stable and a homogenous melt pool were the results of the interaction among double arcs of dual wire. The double-wire technique produced far more age-hardened phases than the single-wire procedure did, which increased the alloy's strength (Figure 3.12) [45].

Materials with such compositions that were challenging to make via the WAAM method with a single-wire feed may now be easily made thanks to the utilisation of multiple feed wires. The dual-wire PAW-based WAAM technique was used to effectively fabricate the intermetallic alloy TiAl [46]. In a related study, the twin wire feed GTAW-based WAAM approach was effectively used in the fabrication of Fe-rich Fe-Al IMCs [47]. The degree of porosity in the final manufactured component is also significantly influenced by the wire's surface quality. In the study mentioned in reference [48], the porosity in the welded connections was minimised by using high-quality wire free of surface scratches and abrasions.

Another important consideration when choosing feed wire in WAAM technology is the nature of the feedstock and the addition of inoculants to it. The nucleation sites via which the grains can nucleate increase as a result of the inoculants' behaviour as heterogeneous nuclei, acting as a possible site for the birth of a large number of grains. The effectiveness of TiN inoculants for efficient grain refining of WAAMed

FIGURE 3.12 Variation in TEM micrograph of age-hardened θ' phase in (a) single feedstock and (b) twin feedstock WAAM [45]. (Open access.)

Ti64 alloy was effectively proven by Kennedy et al. [49]. TiN inoculants encouraged the transformation of columnar grains into equiaxed grains with a reduction in grain size. In WAAM-produced Ti-6Al-4V, thinner–grains with a high equiaxed dendritic morphology were created utilising traces of boron additions [50, 51] and La_2O_3 nucleant particles [52, 53]. Using feedstock wire having components having a strong affinity for O_2, given as Si, can generate oxides in stainless steel WAAM, creating a pinning effect on grain boundaries and limiting grain development [54].

3.8 CONCLUSIONS

Based on the excellent potential and capabilities of wire-fed, arc-based AM techniques, it has attracted the consideration of various industries and manufacturers to use it as a mainstream additive manufacturing technique. A number of parameter selection factors and computer-aided process planning must be taken into consideration in order to determine WAAM's capacity to manufacture almost flawless components made of certain materials. CAD geometry parameters, tool path strategy, the welding method, joining technique parameters, shielding gas, and feedstock selection are some of the main process selection factors covered in this chapter. These process parameters are further detailed with their effect on the mechanical characteristics and microstructure of the fabricated sample.

REFERENCES

[1] B. Wu, Z. Pan, D. Ding, D. Cuiuri, H. Li, J. Xu, J. Norrish, A review of the wire arc additive manufacturing of metals: Properties, defects and quality improvement, *J. Manuf. Process.* 35 (2018) 127–139. https://doi.org/10.1016/j.jmapro.2018.08.001

[2] B. Ralph, Method of making decorative articles. United States, US1533300A. Filed Nov. 12, 1920, Granted April 14, 1925., n.d.

[3] C.R. Cunningham, J.M. Flynn, A. Shokrani, V. Dhokia, S.T. Newman, Invited review article: Strategies and processes for high quality wire arc additive manufacturing, *Addit. Manuf.* 22 (2018) 672–686. https://doi.org/10.1016/j.addma.2018.06.020

[4] S. Pattanayak, S.K. Sahoo, Gas metal arc welding based additive manufacturing—A review, *CIRP J. Manuf. Sci. Technol.* 33 (2021) 398–442. https://doi.org/10.1016/j.cirpj.2021.04.010

[5] P. Dickens, M. Pridham, R. Cobb, I. Gibson, G. Dixon, Rapid prototyping using 3-D welding, in DTIC Document, (1992). https://doi.org/hdl.handle.net/2152/64409

[6] J.D. Spencer, P.M. Dickens, C.M. Wykes, Rapid prototyping of metal parts by three-dimensional welding, *Proc. Inst. Mech. Eng. Part B J. Eng. Manuf.* 212 (1998) 175–182. https://doi.org/10.1243/0954405981515590

[7] B. Tomar, S. Shiva, T. Nath, A review on wire arc additive manufacturing: Processing parameters, defects, quality improvement and recent advances, *Mater. Today Commun.* 31 (2022) 103739. https://doi.org/10.1016/j.mtcomm.2022.103739

[8] B. Tomar, S. Shiva, Microstructure evolution in steel/copper graded deposition prepared using wire arc additive manufacturing, *Mater. Lett.* (2022) 133217. https://doi.org/10.1016/j.matlet.2022.133217

[9] B. Tomar, S. Shiva, Cold metal transfer-based wire arc additive manufacturing, *J. Brazilian Soc. Mech. Sci. Eng.* 45 (2023) 157. https://doi.org/10.1007/s40430-023-04084-2

[10] D. Ding, Z. Pan, D. Cuiuri, H. Li, A practical path planning methodology for wire and arc additive manufacturing of thin-walled structures, *Robot. Comput. Integr. Manuf.* 34 (2015) 8–19. https://doi.org/10.1016/j.rcim.2015.01.003
[11] Z. Pan, D. Ding, B. Wu, D. Cuiuri, H. Li, J. Norrish, Arc welding processes for additive manufacturing: A review, *Trans. Intell. Weld. Manuf.* (2018) 3–24. https://doi.org/10.1007/978-981-10-5355-9_1
[12] P. Singh, D. Dutta, Multi-direction slicing for layered manufacturing, *J. Comput. Inf. Sci. Eng.* 1 (2001) 129–142. https://doi.org/10.1115/1.1375816
[13] D. Jafari, T.H.J. Vaneker, I. Gibson, Wire and arc additive manufacturing: Opportunities and challenges to control the quality and accuracy of manufactured parts, *Mater. Des.* 202 (2021) 109471. https://doi.org/10.1016/j.matdes.2021.109471
[14] M.R. Dunlavey, Efficient polygon-filling algorithms for raster displays, *ACM Trans. Graph.* 2 (1983) 264–273. https://doi.org/10.1145/245.248
[15] S.C. Park, B.K. Choi, Tool-path planning for direction-parallel area milling, *Comput. Des.* 32 (2000) 17–25. https://doi.org/10.1016/S0010-4485(99)00080-9
[16] V.T. Rajan, V. Srinivasan, K.A. Tarabanis, The optimal zigzag direction for filling a two-dimensional region, *Rapid Prototyp. J.* 7 (2001) 231–241. https://doi.org/10.1108/13552540110410431
[17] G.Q. Jin, W.D. Li, L. Gao, An adaptive process planning approach of rapid prototyping and manufacturing, *Robot. Comput. Integr. Manuf.* 29 (2013) 23–38. https://doi.org/10.1016/j.rcim.2012.07.001
[18] D. Ding, Z. Pan, D. Cuiuri, H. Li, A multi-bead overlapping model for robotic wire and arc additive manufacturing (WAAM), *Robot. Comput. Integr. Manuf.* 31 (2015) 101–110. https://doi.org/10.1016/j.rcim.2014.08.008
[19] J.-H. Kao, F.B. Prinz, Optimal Motion Planning for Deposition in Layered Manufacturing, in: Vol. 6 *18th Comput. Eng. Conf., American Society of Mechanical Engineers*, 1998. https://doi.org/10.1115/DETC98/CIE-5699
[20] F. Ren, Y. Sun, D. Guo, Combined reparameterization-based spiral toolpath generation for five-axis sculptured surface machining, *Int. J. Adv. Manuf. Technol.* 40 (2009) 760–768. https://doi.org/10.1007/s00170-008-1385-9
[21] Y. Zhang, Y. Chen, P. Li, A.T. Male, Weld deposition-based rapid prototyping: A preliminary study, *J. Mater. Process. Technol.* 135 (2003) 347–357. https://doi.org/10.1016/S0924-0136(02)00867-1
[22] G.K. Ahiale, Y.-J. Oh, W.-D. Choi, K.-B. Lee, J.-G. Jung, S.W. Nam, Microstructure and fatigue resistance of high strength dual phase steel welded with gas metal arc welding and plasma arc welding processes, *Met. Mater. Int.* 19 (2013) 933–939. https://doi.org/10.1007/s12540-013-5005-3
[23] T. Artaza, A. Suárez, M. Murua, J.C. García, I. Tabernero, A. Lamikiz, Wire arc additive manufacturing of Mn4Ni2CrMo steel: Comparison of mechanical and metallographic properties of PAW and GMAW, *Procedia Manuf.* 41 (2019) 1071–1078. https://doi.org/10.1016/j.promfg.2019.10.035
[24] S.W. Williams, F. Martina, A.C. Addison, J. Ding, G. Pardal, P. Colegrove, Wire + arc additive manufacturing, *Mater. Sci. Technol.* 32 (2016) 641–647. https://doi.org/10.1179/1743284715Y.0000000073
[25] ASTM F3187-16, Standard Guide for Directed Energy Deposition of Metals, ASTM International. West Conshohocken, PA, (2016) 1–22. https://doi.org/10.1520/F3187-16
[26] Q. Wu, J. Lu, C. Liu, X. Shi, Q. Ma, S. Tang, H. Fan, S. Ma, Obtaining uniform deposition with variable wire feeding direction during wire-feed additive manufacturing, *Mater. Manuf. Process.* 32 (2017) 1881–1886. https://doi.org/10.1080/10426914.2017.1364860

[27] B. Wu, D. Ding, Z. Pan, D. Cuiuri, H. Li, J. Han, Z. Fei, Effects of heat accumulation on the arc characteristics and metal transfer behavior in wire arc additive manufacturing of Ti6Al4V, *J. Mater. Process. Technol.* 250 (2017) 304–312. https://doi.org/10.1016/j.jmatprotec.2017.07.037

[28] P. Wang, H. Zhang, H. Zhu, Q. Li, M. Feng, Wire-arc additive manufacturing of AZ31 magnesium alloy fabricated by cold metal transfer heat source: Processing, microstructure, and mechanical behavior, *J. Mater. Process. Technol.* 288 (2021) 116895. https://doi.org/10.1016/j.jmatprotec.2020.116895

[29] K. Pal, S.K. Pal, Effect of pulse parameters on weld quality in pulsed gas metal arc welding: A review, *J. Mater. Eng. Perform.* 20 (2011) 918–931. https://doi.org/10.1007/s11665-010-9717-y

[30] J. Xiong, G. Zhang, W. Zhang, Forming appearance analysis in multi-layer single-pass GMAW-based additive manufacturing, *Int. J. Adv. Manuf. Technol.* 80 (2015) 1767–1776. https://doi.org/10.1007/s00170-015-7112-4

[31] F. Li, S. Chen, J. Shi, Y. Zhao, H. Tian, Thermoelectric cooling-aided bead geometry regulation in wire and arc-based additive manufacturing of thin-walled structures, *Appl. Sci.* 8 (2018) 207. https://doi.org/10.3390/app8020207

[32] G.K. Mohanta, A.K. Senapati, The effect of welding parameters on mild steel by MMAW, *IOP Conf. Ser. Mater. Sci. Eng.* 410 (2018) 012015. https://doi.org/10.1088/1757-899X/410/1/012015

[33] S.P. Tewari, A. Gupta, J. Prakash, Effect of welding parameters on the weldability of material, *Int. J. Eng. Sci. Technol.* 2(4) (2010) 512–516.

[34] A.G. Ortega, L.C. Galvan, S. Rouquette, F.D. Beaume, Effect of welding parameters on the quality of multilayer deposition of aluminum alloy, *Advances in Materials & Processing Technologies Conference*. (2017) Vellore, India, (n.d.). https://doi.org/hal.archives-ouvertes.fr/hal-01909063

[35] J.F. Lancaster, *Metallurgy of welding*, 4th edn (1987) Elsevier.

[36] X. Xu, S. Ganguly, J. Ding, S. Guo, S. Williams, F. Martina, Microstructural evolution and mechanical properties of maraging steel produced by wire + arc additive manufacture process, *Mater. Charact.* 143 (2018) 152–162. https://doi.org/10.1016/j.matchar.2017.12.002

[37] X. Xu, J. Ding, S. Ganguly, C. Diao, S. Williams, Oxide accumulation effects on wire + arc layer-by-layer additive manufacture process, *J. Mater. Process. Technol.* 252 (2018) 739–750. https://doi.org/10.1016/j.jmatprotec.2017.10.030

[38] J. Ding, P. Colegrove, F. Martina, S. Williams, R. Wiktorowicz, M.R. Palt, Development of a laminar flow local shielding device for wire + arc additive manufacture, *J. Mater. Process. Technol.* 226 (2015) 99–105. https://doi.org/10.1016/j.jmatprotec.2015.07.005

[39] P.J. Modenesi, J.H. Nixon, Arc instability phenomena in GMA welding, *Weld. J. Incl. Weld. Res. Suppl.*. 73(9) (1994) 219–224.

[40] C. Matz, G. Wilhelm, Improved arc stability in aluminium welding by oxygen doping of inert shielding gas, *Weld. Int.* 26 (2012) 335–338. https://doi.org/10.1080/09507116.2011.581341

[41] P.L. Miller, K.A. Lyttle, J.B. Neff, D.A. Steyer, K.G. Pierce, Welding gas composition and method for use. (2015) US PATENT 2015/0165565 A1.

[42] L.J. da Silva, F.M. Scotti, D.B. Fernandes, R.P. Reis, A. Scotti, Effect of O2 content in argon-based shielding gas on arc wandering in WAAM of aluminum thin walls, *CIRP J. Manuf. Sci. Technol.* 32 (2021) 338–345. https://doi.org/10.1016/j.cirpj.2021.01.018

[43] P.S. Almeida, S. Williams, Innovative process model of Ti–6Al–4V additive layer manufacturing using cold metal transfer (CMT). In *Proceedings of the twenty-first annual international solid freeform fabrication symposium*. (2010) University of Texas at Austin.

[44] Q. Wu, J. Lu, C. Liu, H. Fan, X. Shi, J. Fu, S. Ma, Effect of molten pool size on microstructure and tensile properties of wire arc additive manufacturing of Ti-6Al-4V alloy, *Materials (Basel)* 10 (2017) 749. https://doi.org/10.3390/ma10070749

[45] S. Wang, H. Gu, W. Wang, C. Li, L. Ren, Z. Wang, Y. Zhai, P. Ma, Study on microstructural and mechanical properties of an Al-Cu-Sn alloy wall deposited by double-wire arc additive manufacturing process, *Materials (Basel)* 13 (2020). https://doi.org/10.3390/ma13010073

[46] L. Wang, Y. Zhang, X. Hua, C. Shen, F. Li, Y. Huang, Y. Ding, Fabrication of γ-TiAl intermetallic alloy using the twin-wire plasma arc additive manufacturing process: Microstructure evolution and mechanical properties, *Mater. Sci. Eng. A.* 812 (2021) 141056. https://doi.org/10.1016/j.msea.2021.141056

[47] C. Shen, Z. Pan, Y. Ma, D. Cuiuri, H. Li, Fabrication of iron-rich Fe–Al intermetallics using the wire-arc additive manufacturing process, *Addit. Manuf.* 7 (2015) 20–26. https://doi.org/10.1016/j.addma.2015.06.001

[48] V.I. Murav'ev, R.F. Krupskii, R.A. Fizulakov, P.G. Demyshev, Effect of the quality of filler wire on the formation of pores in welding of titanium alloys, *Weld. Int.* 22 (2008) 853–858. https://doi.org/10.1080/09507110802650610

[49] J.R. Kennedy, A.E. Davis, A.E. Caballero, S. Williams, E.J. Pickering, P.B. Prangnell, The potential for grain refinement of wire-arc additive manufactured (WAAM) Ti-6Al-4V by ZrN and TiN inoculation, *Addit. Manuf.* 40 (2021) 101928. https://doi.org/10.1016/j.addma.2021.101928

[50] M.J. Bermingham, D. Kent, H. Zhan, D.H. StJohn, M.S. Dargusch, Controlling the microstructure and properties of wire arc additive manufactured Ti–6Al–4V with trace boron additions, *Acta Mater.* 91 (2015) 289–303. https://doi.org/10.1016/j.actamat.2015.03.035

[51] R. Chen, C. Tan, Y. Yu, M. Zhang, X. Yu, C. Liu, W. Ye, S. Hui, Y. Jiang, Modification of α-phase of wire + arc additive manufactured Ti-6Al-4 V alloy with boron addition, *Mater. Charact.* 169 (2020) 110616. https://doi.org/10.1016/j.matchar.2020.110616

[52] M.J. Bermingham, D.H. StJohn, J. Krynen, S. Tedman-Jones, M.S. Dargusch, Promoting the columnar to equiaxed transition and grain refinement of titanium alloys during additive manufacturing, *Acta Mater.* 168 (2019) 261–274. https://doi.org/10.1016/j.actamat.2019.02.020

[53] Y. Chen, C. Yang, C. Fan, Y. Zhuo, S. Lin, C. Chen, Grain refinement of additive manufactured Ti-6.5Al-3.5Mo-1.5Zr-0.3Si titanium alloy by the addition of La_2O_3, *Mater. Lett.* 275 (2020) 128170. https://doi.org/10.1016/j.matlet.2020.128170

[54] F. Yan, W. Xiong, E. Faierson, Grain structure control of additively manufactured metallic materials, *Materials (Basel)* 10 (2017) 1260. https://doi.org/10.3390/ma10111260

4 Wire Arc Additive Manufacturing through GMAW Route

Shazman Nabi, Sandeep Rathee, and M. F. Wani
National Institute of Technology Srinagar, Kashmir, India

Manu Srivastava
PDPM Indian Institute of Information Technology, Design, and Manufacturing Jabalpur, India

4.1 INTRODUCTION

Additive manufacturing (AM) has been in use since the mid-1900s but gained popularity and success in the last few decades because of its several benefits, such as low material wastage, ease of fabrication, design flexibility, and so on [1]. There are several AM approaches in use/research at present, each with its own set of functioning principles. Various committees/researchers have classified them in various ways, for example, the working principle, the physical state of raw material at the time of input and during fabrication, the type of raw material used, and ASTM F42 guidelines [2]. Based on the type of raw material used in the process, AM is divided into two main categories: metal or metallic additive manufacturing (MAM) and non-metallic additive manufacturing. MAM is further divided, based upon the ASTM guidelines, into powder bed fusion (PBF), directed energy deposition (DED), binder jetting (BJ), sheet lamination (SL) techniques, etc. Wire arc additive manufacturing (WAAM) is a variant of the DED technique that employs a heat source in the form of an electric arc to melt the material and deposit it layer by layer to manufacture the component. WAAM has proven to be an appealing technology for the fabrication of medium to large-sized components due to its rapid deposition rate and potentially limitless construction size. It may, for example, make complex shaped components with good precision without the need for traditional machining and cutting tools, fixtures, dies, and other laborious operations in contrast to conventional subtractive processes. Owing to several benefits of WAAM, its usage in various sectors like aerospace, marine, construction, etc., is continuously increasing [3, 4]. WAAM offers greater suitability for different types of materials, such as steel, aluminium, titanium, and magnesium.

Despite its many appealing features, WAAM has a few shortcomings that must be avoided. These include (i) poor component accuracy in terms of dimensions and feature resolution, (ii) residual stress build-up, (iii) heat input distortion, and (iv)

DOI: 10.1201/9781003363415-4

subpar surface finish of the manufactured components. Due to its dependence on arc welding, WAAM requires a significant amount of periodic heat input. As a result, the microstructures of the metal and alloy parts produced by WAAM have complex thermal properties and arc often spatially variable in deposition. Due to the promotion of a large-grain microstructure in WAAM, solidification creates most of the processing problems. Although large grains are advantageous for such systems that require high-temperature creep resistance, they offer less strength, durability, and less resistance to corrosion.

This chapter first briefly classifies WAAM based on the welding techniques utilised and then explains the metal inert gas (MIG)-WAAM in detail. The set-up and working principle of gas metal arc welding (GMAW)-WAAM is then explained. Various process parameters, their effects, and interdependence over each other are defined, and then the various defects incurred during GMAW-WAAM are accordingly elucidated. Various techniques to overcome the defects of MIG-WAAM are also explained in detail. Before concluding the topic, some basic applications of MIG-WAAM are mentioned as well.

4.2 WAAM

WAAM, as mentioned earlier, utilises an electric arc as a source of heat to melt the material/electrode and then deposits it in a layered fashion on the substrate. There are various methods or set-ups for achieving this, viz., gas tungsten arc welding (GTAW)-, plasma arc welding (PAW)-, GMAW-, and hybrid-WAAM (refer Figure 4.1).

The GMAW-based WAAM can achieve an appreciable rate of deposition of about 3 to 4 kgs/hr, which is greater than any other WAAM (arc-based) system. The lowest deposition rate is in the GTAW-based WAAM, followed by the PAW-based WAAM medium. Large components can be produced using tandem GMAW-based WAAM, which uses two heat (power) sources along with a double-wire method to improve the deposition rate. The highest deposition rate possible with a pulse MIG welding process is 9.5 kg/h; however, thick portions must be constructed using an external

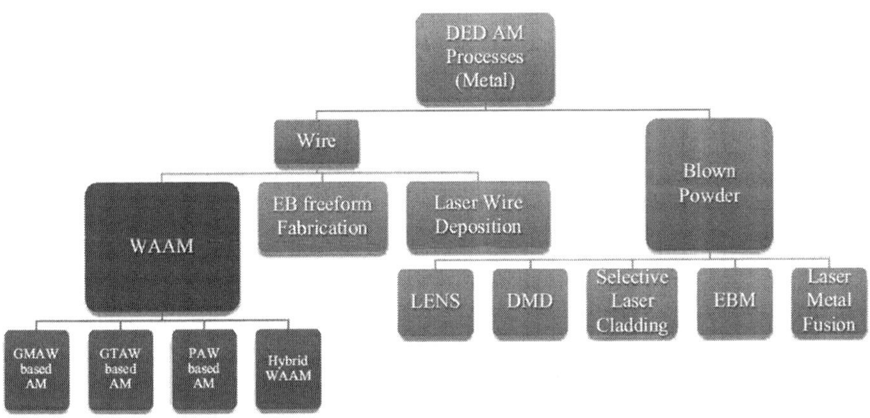

FIGURE 4.1 Classification of WAAM [5].

cooling environment. Moreover, WAAM based on double electrode-GMAW (DE-GMAW) is used to produce metallic goods. It has both a GMAW and a GTAW torch. The GTAW offers a bypass current that is crucial to raising the rate of material usage. With excellent arc stability and low heat input in contrast to the standard WAAM (GMAW-based) processes, cold metal transfer-based (CMT) WAAM has a promising future in AM as one of the most evolutionary variants. With the advantages of substantial quality deposition, minimal distortion, high process tolerance, and nearly no spatter, it is significantly better than the GMAW-based WAAM. The typical rate of deposition might be as high as 2 to 3 kg per hour [6].

It has to be remembered that the previously described WAAM systems use various materials to produce components with various product attributes. The most often used materials in WAAM are steel, titanium alloy, and aluminium alloy. Nickel, copper, and magnesium alloys are also used by some researchers. It is discovered that for the AM of aluminium alloy, GTAW, CMT, and GMAW are all employed, but PAW is not. PAW and GTAW are employed while working with the titanium alloy; however, CMT and GMAW are not. While GMAW is often used in the AM of different steels, the other three arc welding methods are also used in specific steel component manufacturing processes. All four arc welding methods can be employed as the heat source with nickel alloys. When parts are needed in smaller quantities for applications like intricate prototype components and replacement or repair of broken components, the major appeal of GMAW-WAAM for manufacturers is the cheap per unit/part/component cost and minimal production lead time. But before digging deep into MIG/GMAW-WAAM, we must introduce the GMAW process in brief.

4.2.1 GMAW

Among the aforementioned arc welding set-ups, GMAW is the most common of all because most of its equipment is readily available in any welding shop except a few. Due to its greater deposition rates and reduced argon gas consumption cost compared to other arc systems like TIG and plasma systems, MIG is the power source that is most frequently employed for WA-DED of big components.

4.2.2 GMAW Set-Up and Working Principle

Figure 4.2 shows the schematic working principle of GMAW. It uses a consumable wire electrode (the only distinguishable character of GMAW from other MIG/metal active gas (MAG) welding processes) and an inert gas to protect the weld pool from atmospheric contamination. Here is a brief overview of the set-up and its working principle.

4.2.2.1 Set-Up

GMAW set-up varies according to the requirement of work output, but some of the components that are common for all set-ups are briefly defined as follows:

Power source: It uses a constant voltage power source, which provides the electrical energy required to generate an arc between the wire electrode and the workpiece.

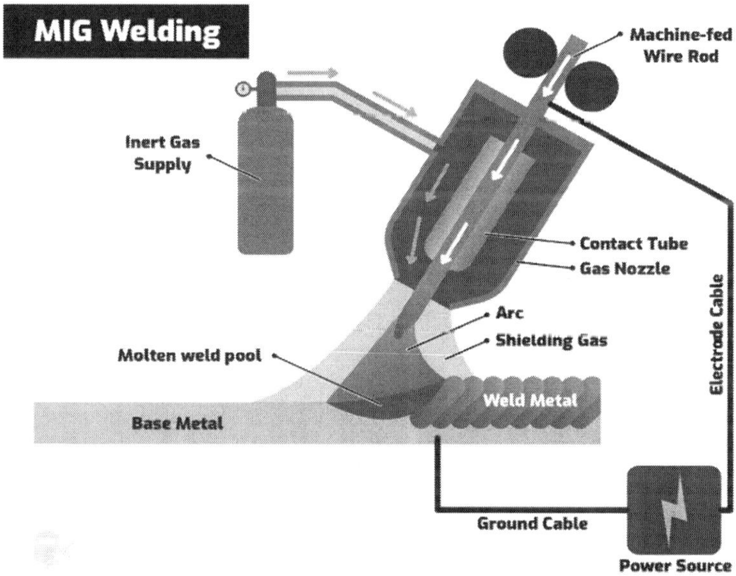

FIGURE 4.2 GMAW working principle [7].

Wire feeder: A spool of welding wire is mounted on the wire feeder. The wire feeder pulls the wire electrode through a cable and delivers it to the welding gun.

Welding gun: It is used to control the feed wire electrode and direct the welding current to the workpiece.

Shielding gas supply: Argon or helium (an inert gas) is supplied through a nozzle on the welding gun to shield the weld pool from any contamination of the surroundings.

Ground clamp: This is connected to the workpiece and completes the electrical circuit required for welding.

4.2.2.2 Working Principle

The welding gun is positioned near the workpiece, and the wire electrode is fed through the gun. When the trigger on the welding gun is pulled, the power source generates an arc between the workpiece and the wire electrode. The heat generated by the arc softens and melts the wire electrode and the workpiece, forming a weld pool. The inert gas flows through the nozzle on the welding gun, creating a shield around the weld pool and protecting it from atmospheric contamination. As the weld pool cools, it solidifies and forms a strong, durable welded joint.

As mentioned earlier, GMAW is a quite cost-effective technique and could be incorporated with AM easily to develop a GMAW-WAAM set-up. When required components are in small quantities for applications like intricate prototype components and for breakdown part replacement or repair, the major appeal of GMAW-WAAM for manufacturers and users is the cheap per part cost and less production lead time. MIG is the most commonly used arc process type in WA-DED.

4.3 GMAW INCORPORATED WAAM PROCESS

A specialised kind of AM for metals known as wire arc DED (WA-DED), commonly referred to as WAAM, is increasingly giving conventional forging and casting techniques a competitive edge. Arc welding equipment that is readily available is employed as the heat source in the WA-DED procedure. As a result, the deposition process parameters have a direct impact on the properties of the layer shape that is deposited. Multiple wires can also be used for MIG with two electrodes and separate power sources to adapt the arc source.

In the MIG/MAG welding process, based on Figure 4.3, a metallic wire electrode is heated by an arc between the metallic substrate and the electrode. The electrode wire melts and then gets deposited on the substrate in the weld pool. A shielding gas is provided around the weld pool and the electrode to prevent oxidation of the component or porosity due to ambient gases and humidity. MAG welding uses an active shielding gas such as carbon dioxide, while MIG welding uses an inert shielding gas such as argon, as mentioned earlier. MIG welding has gained significant commercial success due to its adaptability and low cost in the robotic welding industry.

4.3.1 Various Process Parameters and Their Influence during MIG-WAAM

Given a WAAM hardware system, the fabrication technique employed has a huge impact on the quality of the additively manufactured component/part. It includes extremely complicated multi-physical and interdisciplinary phenomena, and it will still take a lot of work to fully comprehend the physics and material science of this technology and to achieve accurate control over the production process. A number of physical processes, such as heat transmission from the arc created to the electrode wire, melting of wire electrode, creation of droplets, falling and colliding of droplets with the substrate, cooling down and solidification of the beads, and others, are involved and have an impact on the process.

The performance of the finished product and dimensional accuracy may be significantly impacted by each procedure. Many technical factors control these

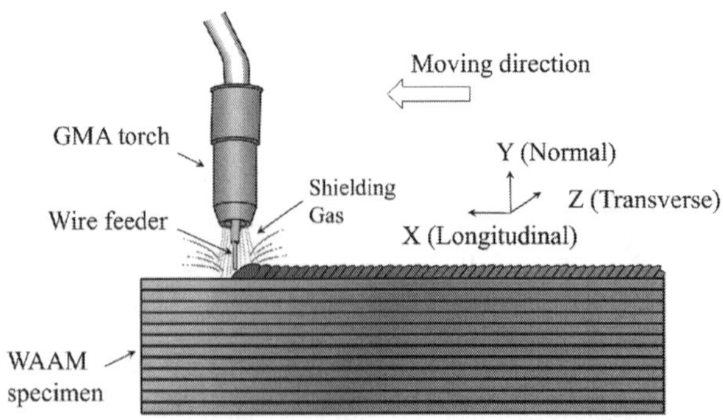

FIGURE 4.3 MIG-WAAM process [8].

operations. Several studies have been conducted to find out how technological factors impact the additively fabricated product's geometry, surface morphology, or mechanical qualities.

4.3.1.1 Input Parameters

There are several input parameters that could affect the geometry, morphology, microstructure evolution, and mechanical properties of the end product of MIG-WAAM. Some of the parameters are current, voltage, heat input, shielding gas flow rate, wire feed rate, dwell time, idle time, etc. Among all the parameters, voltage and current are the most important ones that affect the heat input. Excess heat input can produce a surface that is more in-homogeneous and creates molten pool overflow that leads to lower mechanical properties. Whereas reduction in heat input results in the reduction of mechanical anisotropy. Large currents have the potential to cause instability, overflow, and local collapse. Deposited wall thickness is controlled by one of the most important factors of WAAM, i.e., current. When the current is increased, the height of the deposited layer lowers while the layer width increases. The height of the weld bead decreases while the increase in width is seen with an increase in arc voltage. A thin-walled structure of uniform width can be achieved by controlling the voltage accordingly.

Because a MIG power source lacks arc stability, it limits the capacity to regulate the geometry of the molten pool, which in turn limits the ability to control the geometry of the layers that are deposited [9]. When it comes to large components, the uneven layer geometry results in accumulating geometric errors across the portion. The implementation of the method thus necessitates thorough technological knowledge, which currently, by design standards, is not supported for metal components made using MIG-WAAM.

4.3.1.2 Electrode Polarity

In MIG-WAAM, the polarity of the welding electrode can affect the quality of the weld and the deposition rate. The two types of electrode polarities used in MIG welding are direct current electrode negative (DCEN) and direct current electrode positive (DCEP). DCEN polarity, also known as "straight polarity," is when the electrode is negative and the workpiece is positive. In MIG-WAAM, DCEN polarity is commonly used because it provides better penetration and heat input control, resulting in a stronger and more reliable weld. DCEP polarity, also known as "reverse polarity," is when the electrode is positive and the workpiece is negative. DCEP polarity is generally not used in MIG-WAAM because it can cause a lower deposition rate and a lower quality weld. Around two-thirds of the heat gets produced at the wire electrode, whereas one-third is produced at another terminal, i.e., at the substrate. This is explained by the fluctuation in voltage drop and current density at the cathode, anode, and plasma regions [10]. The arc current stays constant because the same ions and electrons continue to flow between the two electrodes. Since the cross-section of the metal electrode is smaller, current density and heat generation at this terminal are also higher, so it melts rapidly as compared to the higher cross-section of the substrate where the molten electrode is deposited.

The electrode polarity used in MIG-WAAM varies depending on the material being welded and the properties that are desired from the final product. Depending upon the specific welding application, the electrode polarity is chosen.

4.3.1.3 Wire Feed Speed (WFS) and Welding Speed

In MIG-WAAM, WFS is an important process parameter that can affect the quality and efficiency of the deposition process. The amount of material being deposited and the rate at which it is deposited is determined by the WFS. WFS is directly proportional to the deposition rate. As the WFS increases, the amount of material deposited per unit time increases. Therefore, a higher WFS results in greater rates of deposition. It can also affect the shape and quality of the deposited bead. A higher WFS can result in a narrower and taller bead, while a lower WFS can result in a wider and flatter bead. The shape of the bead can also affect the mechanical properties of the deposited material. If the WFS is too high or too low, it can result in some defects, including porosity in the deposited beads. Porosity occurs when gas is trapped in the deposited material, which can weaken its mechanical properties. A higher value can increase the risk of porosity, while a lower value of WFS can result in insufficient material deposition and other defects. It can also affect the heat input to the material being deposited. The microstructure and mechanical properties of the fabricated layers are affected by the higher heat input, which, in turn, is the result of higher WFS. On the contrary, a lower value can result in lower heat inputs, which, in turn, is the reason for incomplete fusion and other defects.

The welding speed is also an important process parameter in MIG-WAAM that can affect the deposition rate, bead shape and quality, porosity and defects, and heat input. It is an important process parameter that can affect the quality and efficiency of the deposition process. It is the rate at which the weld is deposited and can have an impact on the properties of the deposited material. The welding speed is inversely proportional to the deposition rate. As the welding speed increases, the amount of material deposited per unit time decreases. Therefore, a higher welding speed can result in a lower deposition rate. The quality of the deposited bead also depends on the welding speed. A wider and flatter bead profile results from higher welding speeds, while a lower welding speed can result in a narrower and taller bead. The shape of the bead can also affect the mechanical properties of the deposited material. A higher welding speed can increase the risk of porosity, while a lower welding speed can result in incomplete fusion and other defects. The welding speed can affect the heat input as well. A higher welding speed can result in a lower heat input, which can affect the microstructure and mechanical properties of the deposited material. On the other hand, a lower welding speed can result in a higher heat input, which can lead to a higher risk of distortion, cracking, and other defects. Therefore, it is important to optimise the welding speed based on the specific requirements of the application to achieve the desired deposition quality and efficiency.

4.3.1.4 Substrate and Interlayer Temperature

The substrate and interlayer temperatures are also important parameters of WAAM. These process parameters can affect the quality and efficiency of the deposition process. These temperatures are critical because they can affect the important properties (mechanical and microstructural) of the fabricated component. The base material temperature before deposition is called the substrate temperature, and the interlayer temperature is the temperature of the previously deposited layer before the next layer is deposited. The effect of these temperatures in MIG-WAAM can be summarised as follows:

Metallurgical bonding: The substrate and interlayer temperature can affect the metallurgical bonding between the deposited layers and the base material. If the substrate and interlayer temperatures are too low, the deposited material may not bond properly to the base material, leading to poor adhesion and weak mechanical properties. If the temperatures are too high, it can lead to over-tempering or melting of the material, which can also weaken the mechanical properties.

Residual stress: The substrate and interlayer temperatures can also affect the residual stress in the deposited material. If the temperatures are too high, it can lead to higher residual stresses owing to uneven cooling rates. On the other hand, if the temperatures are too low, it can lead to higher residual stresses owing to the difference in thermal expansion coefficients between the deposited material and the base material.

Material properties: The substrate and interlayer temperatures can also affect the material properties of the deposited material, such as hardness, toughness, and ductility. If the temperatures are too high, it can lead to a reduction in material strength and toughness. On the other hand, if the temperatures are too low, it can lead to an increase in material hardness, which can make it more prone to cracking and other defects.

4.3.1.5 Shielding Gas

Shielding gas is used in MIG-WAAM to protect the molten metal from the surrounding atmosphere, which can cause oxidation and other defects in the final product. It is a core part of the GMAW-WAAM process because it aids in shielding or protecting the weld pool. The shielding gas also helps to stabilise the arc and improve the quality of the weld. The majority of gases utilised in GMAW-based AM fall into two categories: active gases and inert gases. Low cost and readily available are two major reasons that carbon dioxide, an active gas, is frequently employed in the GMAW technique (N_2 and H_2 are used in modest increments). Nevertheless, its use promotes the production of oxide, spatter, and deeper penetration (since the presence of O_2 intensifies the localised heat).

Argon (inert gas) is often used as a shielding gas for MIG-WAAM of non-ferrous metals, such as aluminium, copper, and titanium. It provides good arc stability and penetration and helps to prevent oxidation of the molten metal. Helium can be used as a shielding gas for welding high-strength materials, as it provides greater heat input and deeper penetration than argon. Carbon dioxide is a less expensive option that is often used for welding mild steel, but it can produce more spatter and may require higher voltage and amperage settings.

4.4 SOME COMMON DEFECTS ENCOUNTERED IN MIG-WAAM

MIG-WAAM is a relatively new technology for producing metal parts with high geometric complexity and accuracy. However, like any manufacturing process, it suffers from defects that can compromise the quality and functionality of the final product. Some common defects in the MIG-WAAM process include porosity, lack of fusion, cracking, undercut, inclusions, and delamination.

4.4.1 Porosity

Porosity in MIG-WAAM refers to the occurrence of voids within the deposited material, which can compromise some of the properties of fabricated parts. Porosity mainly occurs due to various factors, including improper welding parameters, such as welding speed, current, and voltage, as well as insufficient shielding gas coverage or inadequate cleaning of the workpiece surface. Additionally, the use of low-quality welding wire or improper WFS can also contribute to porosity.

It is essential to optimise the welding parameters and ensure that the welding wire and shielding gas are of high quality to minimise porosity. Proper cleaning of the workpiece surface can also help to eliminate potential contaminants that could contribute to porosity. In some cases, post-processing techniques such as hot isostatic pressing (HIP) or vacuum impregnation may be necessary to further reduce porosity and improve the mechanical properties of the final part.

4.4.2 Lack of Fusion

Lack of fusion is a common welding defect in MIG-WAAM that occurs when the deposited material does not fuse properly with the underlying layer. This can lead to weakened joints and a reduction in the mechanical properties of the final part. There are several factors that can contribute to the lack of fusion, including improper welding parameters such as welding speed, current, and voltage, as well as insufficient preheating or interpass temperature control. Additionally, the use of low-quality welding wire or inadequate shielding gas can also lead to a lack of fusion.

It is important to optimise the welding parameters to ensure adequate heat input and fusion between the deposited material and the underlying layer to address the lack of fusion. Proper preheating and interpass temperature control can also help to reduce the risk of lack of fusion. Additionally, using high-quality welding wire and ensuring adequate shielding gas coverage can help to improve the quality of the deposited material and reduce the risk of lack of fusion. It is also important to inspect the finished part for signs of lack of fusion and other welding defects and to address any issues that are identified through post-processing techniques such as grinding, milling, or re-welding as needed.

4.4.3 Cracking

Cracking is a common defect that can occur during MIG-WAAM and can remarkably compromise the properties, especially mechanical, of the final part. There are several factors that can contribute to cracking, including the following:

Residual stress: During the MIG-WAAM process, residual stress can build up within the deposited material, which can lead to cracking. This is particularly true for large parts or those with complex geometries, which can be more susceptible to thermal stress.

Inadequate interlayer bonding: If the bonding between successive layers is not strong enough, it can create a weak point that is susceptible to cracking. This can occur due to inadequate heat input or poor fusion between the layers.

Inadequate filler material: Using filler material that does not match the base metal can lead to cracking, as the different thermal properties of the two materials can create stress points within the deposited material.

Poor shielding gas coverage: If the shielding gas coverage is inadequate, it can lead to oxidation and porosity within the deposited material, which can create weak points that are susceptible to cracking.

It is important to optimise the welding parameters to minimise residual stress and ensure strong interlayer bonding to address cracking in MIG-WAAM. Proper filler material selection and shielding gas coverage can also help to reduce the risk of cracking. Additionally, heat treatment or stress relief annealing can be used to minimise residual stress and enhance the mechanical properties of the final part. It is also important to inspect the finished part for signs of cracking and address any issues that are identified through appropriate repairs or re-welding as needed.

4.4.4 Undercut

Undercut is a common welding defect that can occur during MIG-WAAM and refers to a groove or depression at the base of the weld that can compromise the mechanical properties of the final part. Undercut can arise because of various factors, such as the following:

Improper welding parameters: If the welding current or voltage is too high or the welding speed is too low, it can lead to excessive heat input and the formation of an undercut.

Inadequate cleaning: If the workpiece surface is not adequately cleaned before welding, it can lead to the formation of contaminants that can contribute to undercut.

Inadequate filler material: Using a filler material that does not match the base metal can lead to the formation of an undercut, as the different thermal properties of the two materials can create stress points within the deposited material.

Poor joint preparation: If the joint is not properly prepared, it can create a weak point that is susceptible to undercut.

It is important to optimise the welding parameters to ensure adequate heat input and fusion between the deposited and the base metal to address undercut in MIG-WAAM. Proper cleaning of the workpiece surface and joint preparation can also help to reduce the risk of undercut. Using high-quality filler material that matches the base metal can also help to reduce the risk of undercut. Additionally, it is important to inspect the finished part for signs of undercut and address any issues that are identified through appropriate repairs or re-welding as needed.

4.4.5 Inclusions

Inclusions are also common during MIG-WAAM and refer to non-metallic materials that are trapped within the deposited material. Inclusions can occur due to several factors, including the following:

Improper welding wire selection: Using welding wire with a high content of non-metallic materials can lead to the formation of inclusions.
Inadequate cleaning: If the workpiece surface is not adequately cleaned before WAAM, it can lead to the formation of contaminants that can contribute to the formation of inclusions.
Poor shielding gas coverage: If the shielding gas coverage is inadequate, it can lead to oxidation and the formation of inclusions within the deposited material.
Improper welding parameters: If the welding current or voltage is too high or the welding speed is too slow, it can lead to excessive heat input and the formation of inclusions.

It is important to use high-quality welding wire and ensure that the workpiece surface is adequately cleaned before welding to address inclusions in MIG-WAAM. Proper shielding gas coverage can also help to reduce the risk of oxidation and the formation of inclusions. It is also important to optimise the welding parameters to ensure that the heat input is sufficient to achieve good fusion between the deposited material and the base metal but not so high as to create excessive melting or the formation of inclusions. Finally, it is important to inspect the finished part for signs of inclusions and address any issues that are identified through appropriate repairs or re-welding as needed.

4.5 GMAW-AM PROCESS CAPABILITY IMPROVEMENT STRATEGIES

Although the GMAW-AM method boasts the greater rates of deposition, lowest production costs, best productivity, maximum energy efficiency, mechanical qualities, and highest density among all other methods, it still suffers from certain shortcomings like anisotropy in properties, long columnar grains, waviness and roughness of surface, etc. Despite these drawbacks, there are various approaches that can be employed to resolve these issues by combining multiple techniques. In order to enhance the properties and characteristics of the produced part, several strategies are adopted alongside the GMAW-AM method. These include path planning, rolling, active cooling, heat treatment processes, surface milling, laser beam machining, passive vision system, dual electrodes, ultrasonic peening, and forging, all of which can further optimise the performance of the GMAW-AM approach.

These common strategies are frequently utilised in combination to improve the overall quality of the final product.

4.5.1 Interpass Cold Rolling

It has been demonstrated that residual strains and deformation may be decreased if the welded bead is rolled within every deposited layer [11]. The implementation of interpass cold rolling can yield more uniform material properties and effectively decrease residual stress. The WAAM method, with its thermal gradient involving deposition layers and alternating re-heating and re-cooling procedures, results in anisotropic microstructural development and mechanical characteristics for the target component. However, by subjecting the deposition to plastic deformation through the cold rolling process, the microstructural anisotropy can be significantly reduced. The interpass cold rolling system schematic diagram is illustrated in Figure 4.4, which was developed at Cranfield University.

By sustaining an external force, a slotted roller is employed to improve the longitudinal tensile strength and refine the microstructure. Interpass cold rolling increases the target component's ultimate tensile strength (UTS) and yield strength (YS) in the direction of build, which results in more uniform material characteristics. In WAAM-fabricated aluminium components, interpass cold rolling can be extremely important in the healing of hydrogen porosity. The rolling process typically results in high dislocation density. These dislocations can serve as 'pipes' for the hydrogen, allowing it to diffuse to the surface, as well as favourable sites for atomic hydrogen absorption.

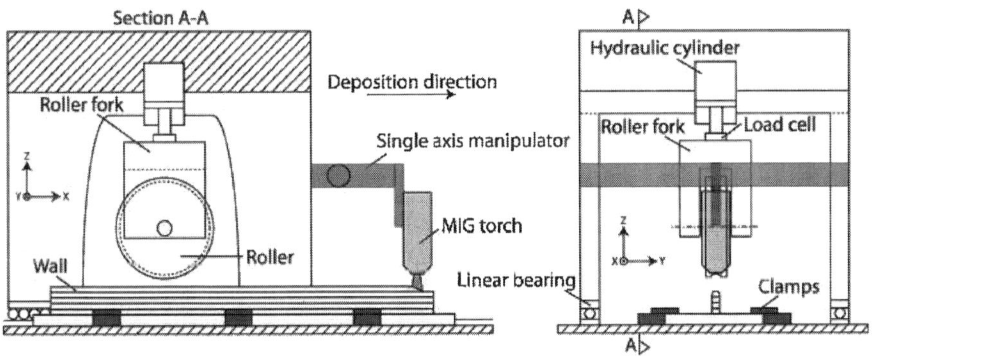

FIGURE 4.4 Schematic diagram of interpass cold rolled MIG-WAAM [12].

4.5.2 Interpass Cooling

High cooling rates result in a more refined microstructure, as is well-documented in the literature. This results in materials that perform better mechanically. They fall into two categories: cooling that is restricted to the weld pool and cooling that affects the entire construction volume. The standard practise is to give each layer a dwell period in between depositions [13]. This makes it possible to control the build-up of too much heat during the deposition process. Yet, as the net deposition rate decreases, this dwell period significantly lengthens the manufacturing time.

Cunningham et al. [14] conducted research on MIG-WAAM and concluded that larger heat input resulted in a faster rate of cooling, which improved the material's tensile strength. Additionally, a higher interpass temperature was found to increase ductility while reducing residual stress and consequential deformation. However, it was observed that a lower weld pool temperature, achieved through the use of both low heat input and low interpass temperature, leads to a more evenly distributed grain structure. This was due to the presence of a significant heat gradient. In contrast, higher temperatures resulted in a more directional heat gradient, which caused an uneven grain structure. Therefore, it is essential to carefully control the heat input and interpass temperature during the deposition process to achieve the desired microstructure, mechanical performance, and geometric characteristics of the deposited material. By adjusting these parameters appropriately, it is possible to optimise the deposition process and obtain high-quality parts with desirable properties.

Subsequent work done by Cunningham et al. focused on the utilisation of liquid nitrogen to rapidly cool down the deposited beads in the weld pool. The distance between the gas exit and the weld power source, as shown in Figure 4.5, was carefully chosen to prevent any disruption to the welding arc. This cooling approach resulted in a much finer, feathery microstructure that more closely resembled the characteristics of a powder-based L-DED process. The WA-DED process, which is typically responsible for columnar grains and anisotropic mechanical behaviour, was effectively interrupted by the cooling mechanism. Due to a change in the highest temperature gradient direction, the nucleation rate increased.

Compared to the approach where the temperature was regulated by an idle time between the depositions of each layer, this cooling method led to an increment in Young's modulus and YS without sacrificing ductility. Additionally, the use of compressed gas, such as CO_2, has been found to further enhance the mechanical performance of Ti-6Al-4V. This is due to the formation of large acicular alpha and refined lamellar alpha grains, which possess more dislocations and grain boundaries, ultimately increasing the material's micro-hardness and UTS.

Furthermore, an air jet impingement approach was employed to accelerate convective cooling and prevent heat build-up during the deposition process. This involved the installation of an air nozzle behind the welding power source to provide quick, localised cooling once the metal was deposited. It was discovered that this technique could be utilised to control the weld pool size and regulate the process's average temperature for each thermal cycle. To further analyse this data, a thermodynamic finite element model was used in conjunction with the experimental findings [13]. Overall, these cooling techniques demonstrate the potential to improve the

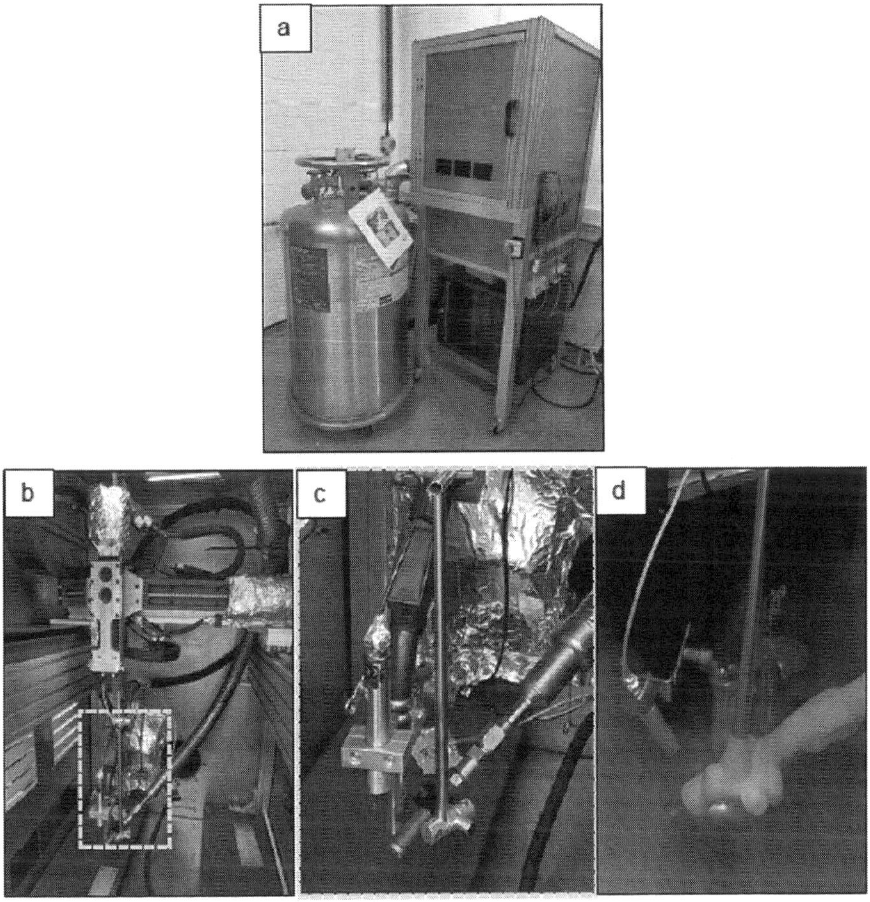

FIGURE 4.5 cryogenic active cooling incorporated WA-DED: (a) machine set up; (b) cryogenic mounted towards z-axis; (c) magnified view of the cryo-nozzle; (d) in process cooling demonstrating the jet of liquid nitrogen [15].

microstructure, mechanical performance, and geometrical features of parts produced through the WAAM process.

4.5.3 Post-Process Heat Treatment

Post-process heat treatment is a crucial step in WAAM to minimise residual stress, enhance material strength, and regulate hardness. The choice of the appropriate heat treatment method is influenced by the target material, working temperature, and heat treatment conditions. Selecting the wrong heat treatment state may increase the likelihood of cracking during mechanical loading since the material's design limit can be exceeded by the combination of residual stress and load stress already present.

Heat treatment has been reported to dramatically increase the mechanical strength of WAAM-fabricated components. Aluminium alloy, intermetallic Ti/Al alloy, nickel-based super-alloys, and titanium alloy have shown increases in strength of

78%, 17%, 5%, and 4%, respectively, following heat treatment [16]. Moreover, post-process heat treatment is essential for grain refinement, particularly for Inconel alloy and aluminium produced using WAAM.

Various heat treatment methods are used in WAAM, including annealing, solution treatment, ageing, quenching, and tempering, to achieve the desired results. Annealing is frequently used to soften the material, reduce residual stress, and increase ductility. Solution treatment and ageing are used to improve the material's strength by precipitating a uniform distribution of fine particles. Quenching is used to increase the material's hardness, and tempering is used to decrease brittleness and increase toughness.

The heat treatment process may be influenced by several factors, including the material's composition, deposition speed, and thickness. The heat treatment time and temperature must be carefully controlled to avoid material damage or cracking. Post-process heat treatment is critical for attaining the desired mechanical properties in WAAM-fabricated components, and it must be tailored to the specific material and application. The material alloying system and the preheat treatment condition both influence the decision to apply post-process heat treatment. Most materials with a high carbon concentration must be heat treated, while certain materials may be harmed by this process. Thus, the individual material and its application must be considered when using the post-heat treatment technique on a WAAM component.

4.6 APPLICATIONS OF GMAW-WAAM

It is evident that the MIG-WAAM process is highly suitable for manufacturing large-sized parts and components with numerous components, whether intricate or of large scale, having been already manufactured through this process. Some of the broad applications of MIG-WAAM are as follows:

Aerospace: MIG-WAAM can be used to manufacture aerospace components, such as turbine blades, landing gear, and engine parts. The technology allows for the creation of complex geometries and the use of high-performance materials.

Automotive: It can be used to manufacture automotive components, such as engine blocks, transmission components, and suspension parts. The technology offers the potential for lighter-weight components with improved strength and durability.

Tooling: It can be used to manufacture tooling components, such as moulds, dies, and fixtures. The technology allows for the creation of customised tooling with reduced lead times and lower costs.

Architecture: It can be used to manufacture architectural components, such as building facades and structural elements. The technology offers the potential for the creation of unique designs with reduced material waste.

Overall, MIG-WAAM has the potential to be used in a wide range of industries and applications where the production of large-sized, complex, and intricate metal parts is required.

4.7 CONCLUSION AND FUTURE PROSPECTS

The prospects of MIG-WAAM look promising, as this technology offers several advantages over traditional manufacturing processes. Some of these advantages include faster production times, reduced material waste, the ability to manufacture complex geometries, and the potential to use a wider range of materials.

MIG-WAAM technology has already been adopted in various industries, such as aerospace, automotive, and defence. However, there is still room for growth and development in the technology, particularly in terms of process optimisation and material selection.

In conclusion, MIG-WAAM is a technology with significant potential in the manufacturing industry. As research and development continue to improve the technology, it is likely that we will see increased adoption and application of MIG-WAAM in various industries, leading to improved efficiency and reduced costs.

REFERENCES

1. Srivastava, M., Rathee, S., Maheshwari, S., Kundra, T. K., *Additive Manufacturing: Fundamentals and Advancements*. 2019, Boca Raton, FL: CRC Press.
2. Srivastava, M., Rathee, S., Additive manufacturing: Recent trends, applications and future outlooks. *Progress in Additive Manufacturing*, 2022. **7**(2): pp. 261–287.
3. Srivastava, M., et al., Wire arc additive manufacturing of metals: A review on processes, materials and their behaviour. *Materials Chemistry and Physics*, 2023. **294**: p. 126988.
4. Srivastava, M., et al., A review of various materials for additive manufacturing: Recent trends and processing issues. *Journal of Materials Research and Technology*, 2022. **21**: pp. 2612–2641.
5. Srivastava, S., Garg, R. K., Sharma, V. S., Sachdeva, A., Measurement and mitigation of residual stress in wire-arc additive manufacturing: A review of macro-scale continuum modelling approach. *Archives of Computational Methods in Engineering*, 2021. **28**(5): pp. 3491–3515.
6. Li, Y., Su, C., Zhu, J., Comprehensive review of wire arc additive manufacturing: Hardware system, physical process, monitoring, property characterization, application and future prospects. *Results in Engineering*, 2022. **13**: p. 100330.
7. Weldguru, *What is GMAW (MIG Welding) & How Does it Work?* Sep. 1, 2022; Available from: https://weldguru.com/mig-welding/
8. Abusalma, H., Eisazadeh, H., Hejripour, F., Bunn, J., Aidun, D. K., Parametric study of residual stress formation in wire and arc additive manufacturing. *Journal of Manufacturing Processes*, 2022. **75**: pp. 863–876.
9. Zhang, Z., Sun, C., Xu, X., Liu, L., Surface quality and forming characteristics of thin-wall aluminium alloy parts manufactured by laser assisted MIG arc additive manufacturing. *International Journal of Lightweight Materials and Manufacture*, 2018. **1**(2): pp. 89–95.
10. Pattanayak, S., Sahoo, S. K., Gas metal arc welding based additive manufacturing—A review. *CIRP Journal of Manufacturing Science and Technology*, 2021. **33**: pp. 398–442.
11. Colegrove, P.A., Donoghue, J., Martina, F., Gu, J., Prangnell, P., Hönnige, J., Application of bulk deformation methods for microstructural and material property improvement and residual stress and distortion control in additively manufactured components. *Scripta Materialia*, 2017. **135**: pp. 111–118.

12. Colegrove, P.A., Coules, H. E., Fairman, J., Martina, F., Kashoob, T., Mamash, H., Cozzolino, L. D., Microstructure and residual stress improvement in wire and arc additively manufactured parts through high-pressure rolling. *Journal of Materials Processing Technology*, 2013. **213**(10): pp. 1782–1791.
13. Montevecchi, F., Venturini, G., Grossi, N., Scippa, A., Campatelli, G., Heat accumulation prevention in wire-arc-additive-manufacturing using air jet impingement. *Manufacturing Letters*, 2018. **17**: pp. 14–18.
14. Cunningham, C.R., Wang, J., Dhokia, V., Shrokani, A., Newman, S.T., Characterisation of Austenitic 316 LSi Stainless Steel Produced by Wire Arc Additive Manufacturing with Interlayer Cooling, in *2019 International Solid Freeform Fabrication Symposium*, 2019, University of Texas, Austin, TX.
15. Cunningham, C., Pulse Metal Inert Gas based Wire Arc Additive Manufacturing of an Austenitic Stainless Steel. Doctoral dissertation, University of Bath, 2020.
16. Baufeld, B., Brandl, E., van der Biest, O., Wire based additive layer manufacturing: Comparison of microstructure and mechanical properties of Ti–6Al–4V components fabricated by laser-beam deposition and shaped metal deposition. *Journal of Materials Processing Technology*, 2011. **211**(6): pp. 1146–1158.

5 Cold Metal Transfer-Based Wire and Arc Additive Manufacturing (CMT-WAAM)

Bunty Tomar and S. Shiva
Indian Institute of Technology Jammu, India

5.1 INTRODUCTION

The ability of wire and arc-based additive manufacturing (WAAM) to manufacture bulk and big components by the high rate of material deposition, effective raw material utilisation, lesser infrastructure cost, and consequent environmental friendliness has attracted the industrial manufacturing sector more and more in recent years [1]. The patent of WAAM was filed by Ralph in 1925 when he used an electric arc as a source of fusion energy along with metallic wire for the raw material for creating metallic decorations [2]. In this technology, the wire-feeding mechanism and the arc welding principle are integrated. By utilising the provided 3D CAD model data, the wired feed material is melted employing the heat of the created arc, and then molten metal is principally placed on a specified substrate in a layer-upon-layer manner. A robot-assisted arm and gantry system machine uses this continuous layer-upon-layer deposition to manufacture the entire 3D metal structure [3].

The WAAM method can be broadly divided into three categories based on the kind of thermal energy source used: gas metal arc welding (GMAW) [4], gas tungsten arc welding (GTAW) [5], and plasma arc welding (PAW) [6]. The primary determining element in choosing the precise welding source for a given WAAM application is the user requirement. If the application demands a greater deposition rate, GMAW-based WAAM could be utilised instead of GTAW-based WAAM. For tracks with less distortion, PAW-WAAM should be preferred over GMAW-WAAM. The choice of welding methods in the WAAM technique affects the amount of time required, the rate of deposition, and the necessary processing parameters for a particular alloy [7]. Reference [7] presents a summarised detail of all these types of WAAM techniques. Cold metal transfer-based wire arc additive manufacturing (CMT-WAAM) is another variant of WAAM-based on GMAW-WAAM and is known to excel other contemporary WAAM techniques in means of controlled heat input and spatter-less deposition [8].

The research articles that have been published on the application of CMT-based wire and arc additive manufacturing are critically reviewed in this chapter. This chapter's material includes an overview of CMT-WAAM, as well as information on

DOI: 10.1201/9781003363415-5

the CMT process's precise mechanism and advantages over other WAAM variations. The shown applications of CMT-WAAM in several materials, consisting of various grades of steel, nickel super alloys, titanium and its alloys, and aluminium alloys, are compiled with the usual microstructure and mechanical properties found. In the end, some of the recent advances in the field of CMT-WAAM are followed by a concluding discussion in such a track that it will provide guidance to beginners in the field of CMT-WAAM.

5.2 COLD METAL TRANSFER (CMT)

CMT is a modified variant of the GMAW technology and was established by Fronius in the early 2000s. The CMT process is based on the short-circuiting of the arc current and differs from the GMAW technique in terms of droplet detachment from the feedstock wire [9]. There are three ways to transport molten metal in the GMAW-WAAM: spray, globular, and short-circuiting [10]. Small droplets move over the arc in a spray mode of transfer at higher currents and under the influence of electromagnetic fields. Despite the fact that drop detachment is spatter-free, its application is limited by greater heat inputs. The globular transfer happens at lower currents and involves the movement of discrete metal droplets of the size of diameter of the feedstock wire within the arc gap as a result of gravitational force. Spatter is a common side effect of this phenomenon due to the uneven globular transfer. In short-circuiting transfer, the molten metal at the wire's tip is carried from the feedstock to the molten pool when it comes into contact with the pool surface, causing short-circuiting and aiding the spatter-less transfer. The smallest welding currents are used during short-circuiting transfer, which produces a tiny weld pool [11]. The phenomenon of short-circuiting with lesser welding currents is employed in the CMT-based WAAM technique as the mode of metal transfer.

5.2.1 CMT Process

The oscillating feedstock machinery and high-speed digitally controlled system initially created for CMT welding are used in CMT-WAAM [12]. The crucial component of this process is the retraction of the wire while transferring the material. As soon as the tip of the feedstock reaches the melt pool, a digitally controlled procedure causes the welding torch's associated servomotor to reverse. This method enhances the transfer of molten metal droplets without using electromagnetic force by retracting the feedstock wire. CMT uses less current than traditional GMAW to transmit the same quantity of material. The need for a larger current for a given material transfer is cancelled by the retraction force [13]. A major contributor to low heat input is oscillating wire movement, which gives molten wire more kinematic energy for easier transfer into the melt pool. While using CMT, spatter production is also eliminated by lowering the current to almost nil just before transferring molten metal. Once the metal is transferred, the current is increased, the arc gets re-ignited, while the feedstock is pushed forward by the welding torch's servo metre [9].

The deposit of molten feedstock wire droplets in the melt pool completes one electrical cycle for the CMT process. Some significant elements affecting the regulation of the spatter and the size of the generated drop and its detachment are the

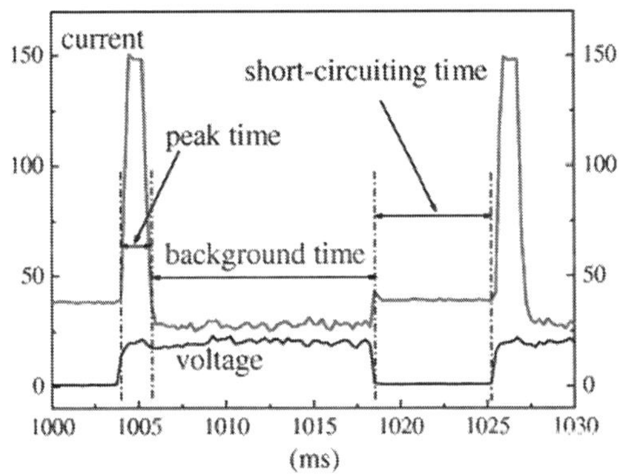

FIGURE 5.1 Electrical and voltage cycle waveforms for the CMT process [13]. (Reprinted by permission of Elsevier.)

magnitudes of the current and voltage, the wire-feed speed, the speed of travel, and the frequency [14]. The CMT electrical cycle waveforms (seen in Figure 5.1) are divided into three time zones as follows:

The peak current zone: The welding arc easily erupts during this phase due to a persistently high current and persistent arc voltage. The generated arc heats the feedstock wire and melts it, creating liquid droplets.

The background current zone: At this time, which has a comparatively lower current and steady voltage, happens soon before the short-circuiting. Lower current intensities stop the globular transfer of the metal droplets in the molten pool.

The short-circuiting zone: Zero arc voltage is provided in this phase. The wire-feeding servomotor receives the reverse signal, which causes it to pull the wire to the opposite side of the melt pool. This event aids in the droplet's separation from the feedstock tip and transport into the molten pool [13, 15].

5.2.2 Comparison of CMT with Other Traditional MIG Methods

The CMT process's technology has twin key components: first, a steady and regulated short-circuiting that produces less spatter, and, second, a lesser heat input given by a lowered current at the time of short-circuiting due to the comparatively low short-circuiting current, CMT results in a smaller droplet generation and subsequent steady separation of the liquid bridge among the feedstock and molten pool. As the traditional MIG method allows for a significant short circuit current, electric explosions happen during metal transfer and detachment of droplets. It results in the production of small amounts of fine spatter and relatively large heat inputs [16]. As a result, CMT has a number of advantages over the traditional MIG technique, including minimal energy input and spatter-free operation. Panchenko et al. [17] demonstrated that in comparison to GMAW-WAAM with the same processing conditions, CMT-WAAM of an Al-based alloy displayed a 16% reduced heat input.

In wire and arc AM, increased heat input is a cause for concern since too much heat during deposition causes previously deposited layers to remelt to a greater extent, degrading the macrostructure and bead shape. The formation of larger residual strains and distortions is also a worry when there is a higher heat input. A uniform deposition surface is provided with a reduced risk of the weld collapsing and overflowing with a lower equivalent heat input [18]. By limiting the gas solubility amount inside the molten pool and by controlling the temperature of liquid droplets, lesser heat input is also advantageous for lowering the combined porosity inside the manufactured component, particularly in Al-based alloys [19]. Hence, CMT-WAAM's low heat input is essential for avoiding the negative consequences of a larger heat input. CMT-WAAM generates more reproducible bead shape and size for thin wall and bulk samples than standard MIG-based WAAM [20]. A schematic of CMT-WAAM is presented in Figure 5.2.

With the exception of the MIG technique, CMT outperforms GTAW and PAW WAAMs on the basis of total heat incurred, spatter production, beginning costs, wire-feeding complexity, and rate of metal deposition. When compared to CMT, GTAW, and PAW-based processes are more challenging to operate because they require separate, non-coaxial wire-feeding systems. With no need to modify the wire-feeding mechanism, GMAW-WAAM (including CMT) methods outperform GTAW and PAW techniques in the case of multidirectional deposition paths. The advantages of CMT-WAAM over the other wire and arc AM versions, such as its reduced heat input requirements, ability to regulate spatter, and greater deposition rate, may, therefore, be considered justifiable. Figure 5.3's spider diagram compares CMT with GTAW and PAW methods in light of the research review. In the spider diagram, where one represents the least and five the most, comparisons are made on the basis of deposition rate, given process heat input, spatter formed during operation, ease in feeding mechanism, and capital spent in machinery establishment.

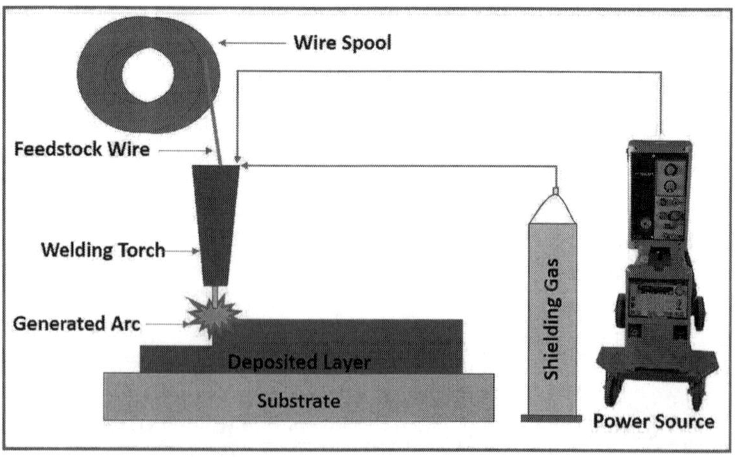

FIGURE 5.2 Schematic representation of the CMT-WAAM process.

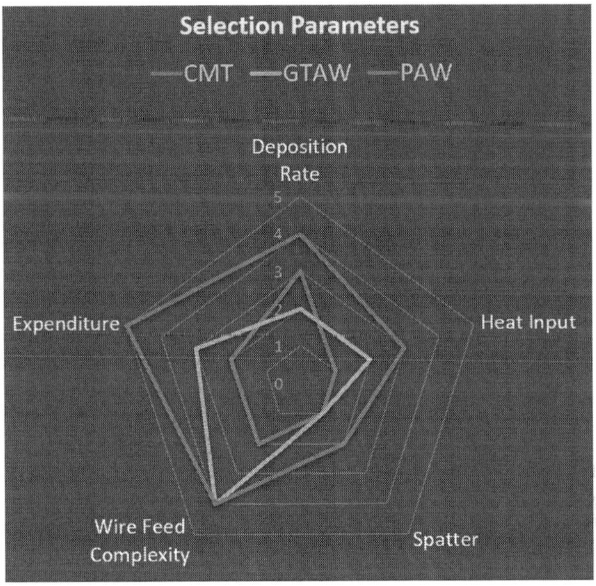

FIGURE 5.3 Schematic comparison of the CMT process with GTAW and PAW process [8]. (Reprinted by permission of Springer.)

5.3 ALLOYS USED IN CMT-WAAM

For manufacturing using the CMT-WAAM method, spooled wires of different alloys that are commercially available are employed. The majority of these wires are produced for use as feedstock electrodes in the joining and welding industries. A thorough awareness of current technique alternatives, process management strategies, and the right feed material for a certain application is necessary for the production of a defect-free and dependable component. In this portion of the chapter, the commonly utilised metals with their alloys in the CMT-WAAM method are reviewed, with an emphasis on the microstructural and mechanical characteristics of the components produced.

5.3.1 Aluminium Alloys

The typical processing parameters are the main determinants impacting the characteristics of deposited structures for CMT-WAAM of aluminum-based alloys. In their research, Wang et al. [21] showed that it was possible to modify the process parameters, including the wire-feed rate and the short-circuiting current, to control the energy input. The surface quality of the feedstock material has a significant impact on the production of porosity in contrast to the bulk composition of feedstock and its feeding speed, CMT modes, and torch travel [22]. The production of Al alloys via CMT-WAAM involves a sizeable amount of residual stress. The produced stresses are significantly influenced by the substrate's size and the height attained by the deposited structure. The magnitude of residual stresses in deposited aluminum components is also significantly influenced by the implemented clamping mechanism [23]. The deposition geometry of aluminum alloys produced by CMT-WAAM is also a result of the distinctive substrate factors, such as the size dimensions and starting substrate temperature [24].

ER2319 alloy is a crucial aluminium alloy for fabrication using the WAAM approach because it has outstanding mechanical qualities and weldability. Mechanical characteristics were isotropic in the as-fabricated Er2319 alloy, according to Dong et al. [25], which was caused by the reduced Cu concentration in α-Al and the fact that fractures developed along thick, net-like secondary phases without any directional growth after nucleation. Iron-rich phases did not have any significant impact on the beginning and development of fractures in interlayer areas. Mechanical characteristics showed post-heat treatment anisotropy because α-aluminum strengthened and obtained increased resistance for fracture initiation. Fracture opening towards the direction of the construction was connected towards concentrated iron-rich phases in interlayer areas. The interlayer zones were also perpendicular to the direction of construction and the tensile loading force.

5.3.2 Steels

Steels are recognised for having strong mechanical properties and exceptional ductility, which make them appropriate for a variety of industrial applications. Stainless steels offer good corrosion resistance in a variety of environmental situations. As a result, several studies of the production of stainless steels [26, 27], tool steels [28, 29], high-strength steels, bimetallic steels [30], and high-strength, low-alloyed steels are being conducted via the CMT-WAAM route.

Peers generally agree that the product's microstructure is influenced by the manufacturing process's temperature cycle. The degree of cooling has a significant impact on the mechanical strength and development of the microstructure in stainless steel [31]. Thus, in CMT-WAAM, it is crucial to establish the substrate's pre-heat temperature and pre-heat time. Ge et al. [32] observed that the ultimate tensile strength (UTS) of the lower and central layers grew gradually with a fast rise towards the top layers in CMT-WAAM fabricated geometries of 2Cr13 steel (shown in Figure 5.4). It is due to the slowing down of cooling in the layers below and the progressive build-up of martensite from the centre to the top layers. Figure 5.5 depicts the distinctive optical microstructure of this report.

The selection of inappropriate processing parameters during steel manufacturing results in the formation of anisotropic mechanical characteristics and deposit porosity [33]. These properties are largely dependent on the microstructure and texture of the deposit [34]. In general, austenite in stainless steels comes in a variety of morphologies, named intragranular, Widmanstätten, and grain boundary γFe with a respectable quantity of αFe. The repeated heating and cooling series in CMT-WAAM have a significant impact on the existence of austeno-ferrite phases [35].

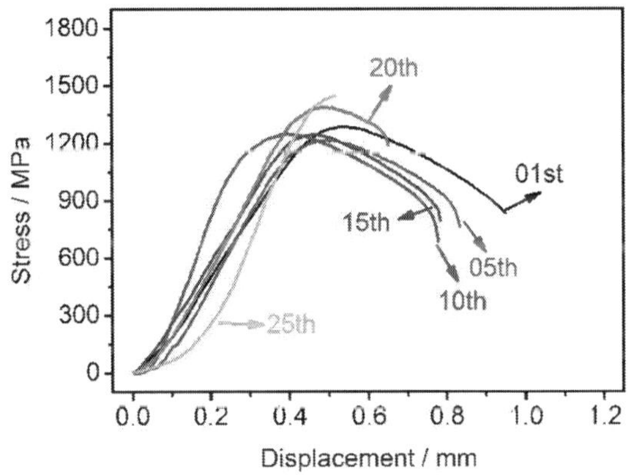

FIGURE 5.4 Stress displacement graphs for the marked layers [32]. (Reprinted by permission of Elsevier.)

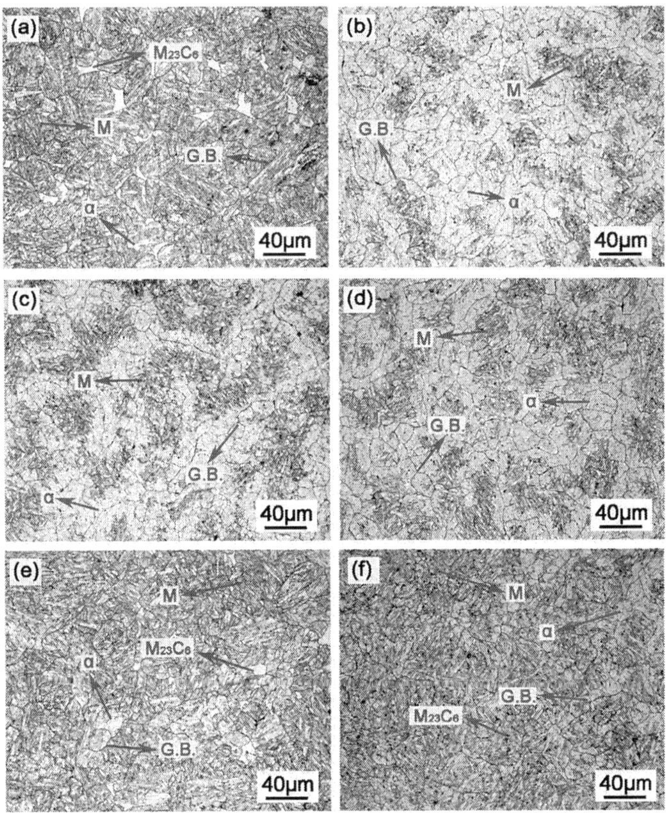

FIGURE 5.5 Corresponding optical micrographs for (a) 1st, (b) 5th, (c) 10th, (d) 15th, (e) 20th, (f) 25th layers [32]. (Reprinted by permission of Elsevier.)

5.3.3 Nickel-Based Alloys

Because of the extremely high heat input and intricate thermal sequences needed during WAAM, nickel-based superalloys produced by this method frequently have poorer component quality (including the existence of fractures and excessive surface roughness) [36]. With the precipitation of the Laves phase and MC carbides inside the inter-dendritic region, Inconel 718 alloy produced by WAAM exhibits a microstructure made up of upwards-oriented columnar grains [37]. The Inconel 718 alloy created by WAAM has a microstructure made up of upwardly oriented columnar grains having the precipitation of the Laves phase and MC carbides at the inter-dendritic region [38]. Figure 5.6 depicts the change in the microstructure of the IN718 alloy by changing the travel speed, which includes the variation in

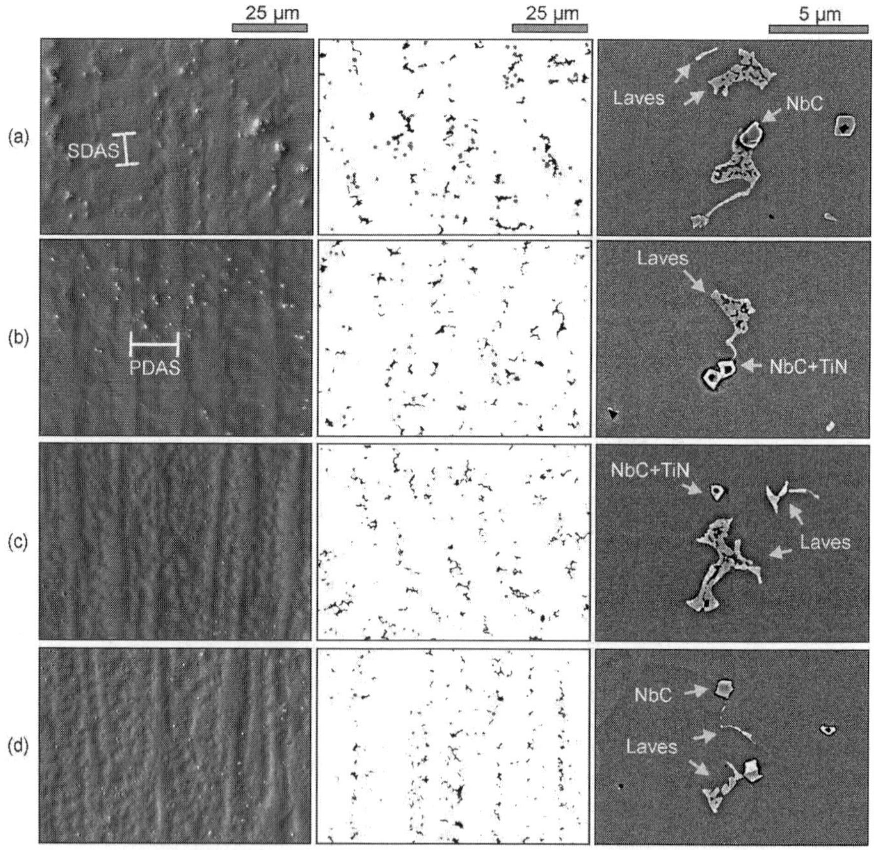

FIGURE 5.6 Microstructural variation on varying travel speed in IN718 alloy: (a) 0.2 m/min, (b) 0.4 m/min, (c) 0.6 m/min, (d) 0.8 m/min [38]. (Open access.)

geometry of the carbide and Laves phases. The significant heat intensities exacerbate the enrichment of alloying elements like molybdenum and niobium, cause the creation of unfavourable Laves phases, and have a negative impact on the microstructure and mechanical properties [39]. CMT-WAAM has become a viable solution to deal with these heat-related problems. By regulating the cooling rates, in situ heat treatment of IN718 alloy during CMT-WAAM deposition is possible because of the production of strengthening phases such as the δ, γ' and γ'' phases, such heat treatment results in an ageing treatment [40]. Due to the impact of the precipitated strengthening stages, the deposit's microhardness increased from 250 HV to 305 HV.

5.3.4 Titanium-Based Alloys

The mainstream research on Ti-based alloys is being done on Ti-6Al-4V alloy [41, 42]. Based on the pace of cooling from the transus temperature, Ti-6Al-4V material exhibits a variety of microstructural characteristics. With slower cooling rates, the phase initially forms at the grain boundaries (αGB) surrounding the previously formed β grains. Following the generation of the αGB, primary α deposits in parallel lamellae, producing clusters that are perpendicular to the αGB. At the time of the faster cooling rate within the grains, primary α is solidified in the basketweave structure. At an exceptionally rapid cooling rate (greater than 525°C/s), non-diffusional reactions produce martensite (α') [43]. Findings indicate that CMT-WAAM of the TI6Al4V is challenging because it frequently experiences arc instability and excessive spatter generation [44]. A crucial element in determining the bead shape is the produced arc force during deposition [45]. Another element affecting the geometry of the deposited structure is the size and spread capability of the melt pool. The amount of oxide skin that forms over the molten pool during solidification has an impact on the deposit's shape as well [46]. Using a high-speed camera, Zhou et al. studied the material transfer behaviour in CMT-WAAM of Ti-6Al-4V and reported that the production of stable arc and spatter-less deposition is possible at currents (I) of 300 Ampere.

The mechanical characteristics of Ti-64 alloy samples fabricated by CMT-WAAM are in good agreement with the wrought Ti-64 [47] and superior to that of as-casted Ti-6Al-4V [48] samples. These samples have anisotropic characteristics in their as-manufactured state, which is primarily due to their microstructure, which contains α lamellae with elongated precursor β grains. Vazquez et al. [49] used CMT-WAAM to create a wall of Ti-64 alloy that was flat and uniform without any defects, and they then used the proper post-fabrication heat treatments to reduce the observed mechanical anisotropy and enhance the ductility of the raw materials. Different heat treatments result in a noticeable change in mechanical characteristics and microstructure (as seen in Figures 5.7 and 5.8). Hybrid CMT-based WAAM may present an opportunity to improve weld stability and boost deposition rate because it has produced promising results in its preliminary study [50].

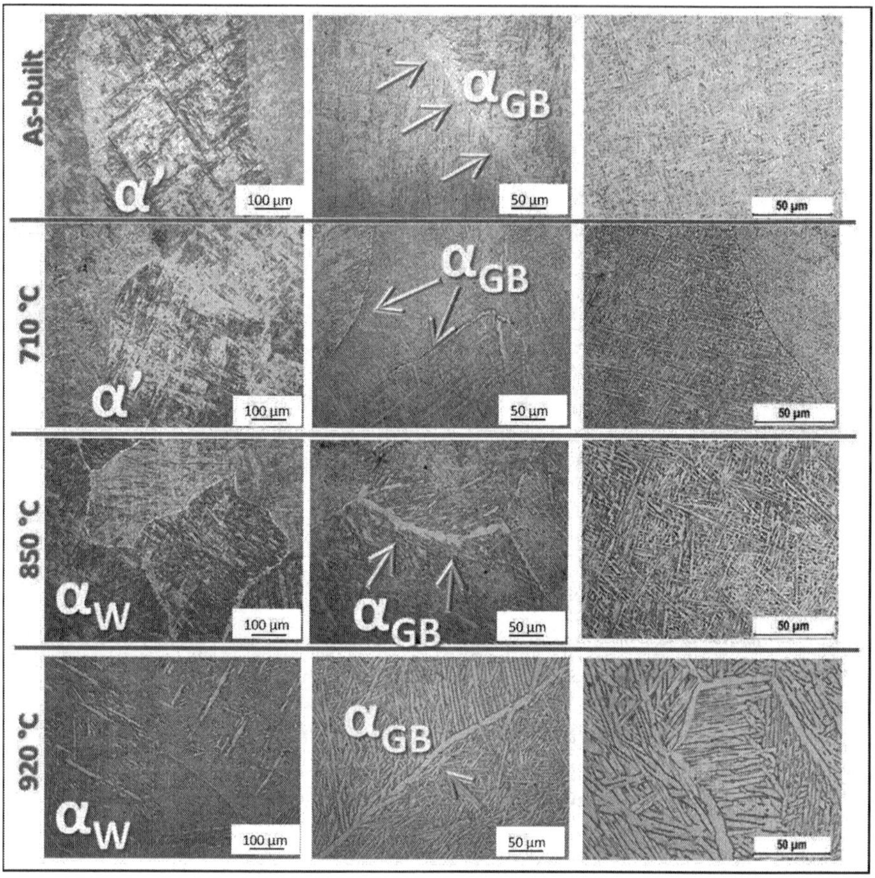

FIGURE 5.7 Microstructure evolution in CMT-WAAM fabricated Ti64 alloy in as-fabricated and heat-treated (HT) conditions (α': martensite; α_w: Widmanstätten α; α_{GB}: α at the grain boundary) [49]. (Open access.)

5.4 RECENT ADVANCES IN CMT-WAAM

The main fabrication quality indicators of CMT-coupled WAAM, coupled with the process parameters, are the distinctive solidification path and thermal history. The determining factors to effectively make high-quality CMT-WAAM components include the key process selection, geometry-integrated heat dissipation rate, and fabrication method. The final functionality of the deposited structure is also affected by the thermal gradient. These elements regulate the component's quality and anticipated heterogeneity while it is being created [51]. Several steps are being taken, and advancements are being made in the tool assembly, operation, and process of CMT-WAAM to reduce the distortions and negative consequences caused by high heat input. A summary of several of these recent developments in the field of CMT-WAAM is included in this portion of the chapter.

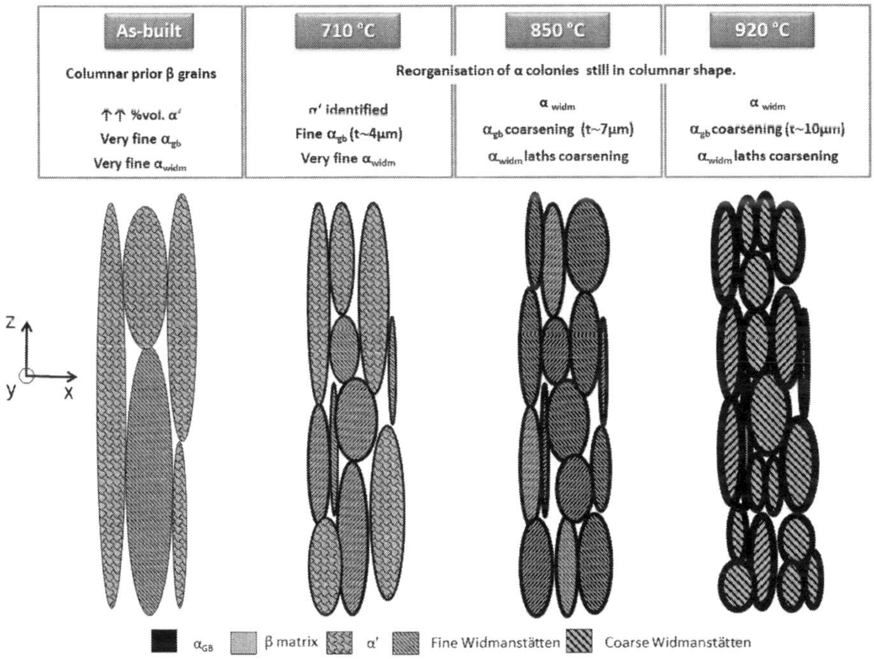

FIGURE 5.8 Schematic illustration of microstructure evolution in CMT-WAAM fabricated Ti64 alloy in as-manufactured and HT conditions [49]. (Open access.)

5.4.1 Near Immersion Active Cooling (NIAC)

The major difficulty in WAAM-based applications is thermal control. The continuous cyclic heating and cooling process, along with the interpass temperature encountered while deposition, reflect a significant influence over the microstructure and mechanical characteristics of a WAAM deposited structure as they are formed [52]. In the WAAM approach, NIAC is newly created and tested. NIAC aims to improve the process of thermal management in the wire and arc AM to reduce heat build-up brought on by excessive heat input. The NIAC concept was presented by Silva et al. [53] based on the active cooling of the previously formed structure by immersing the deposit in a cooling liquid in a prescribed and controlled fashion.

By reducing the build-up of heat during the WAAM manufacturing of the AWS ER5356 alloy, Silva et al. [54] confirmed the usage of the NIAC for their research. The experimental set-up for the NIAC idea is depicted in Figure 5.9b. Figure 5.9 depicts three distinct methods of doing the deposition that were used to demonstrate the viability of the NIAC idea (b). In comparison to the other two techniques, the outcomes indicated that NIAC-deposited walls had with highest geometric tolerance and had the least proportion of voids and porosity. It has been demonstrated that this technique offers enhanced mechanical characteristics with lesser anisotropy. Scotti et al. [55] published similar results for thermal management in CMT-WAAM manufactured structures employing the discussed NIAC approach.

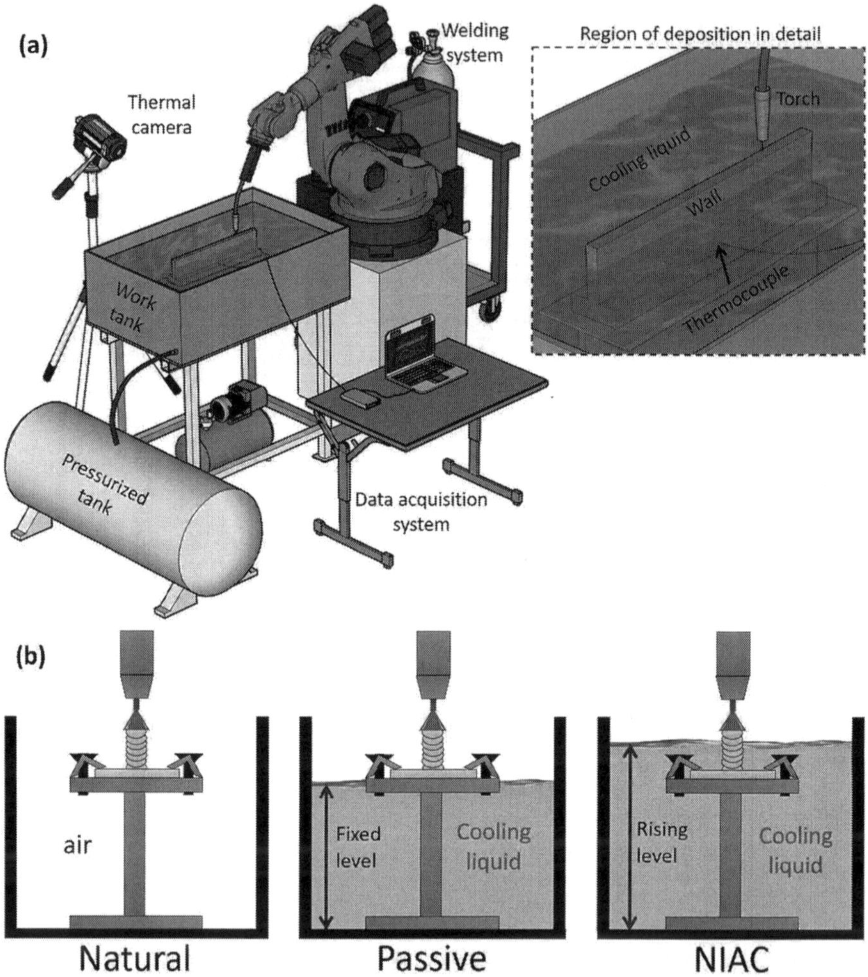

FIGURE 5.9 (a) Schematic experimental set-up of NIAC approach; (b) cooling approaches adopted [54]. (Reprinted by permission of Springer.)

5.4.2 CMT Pulse Advanced (CMT-PADV)

Porosity has been noted in the literature as a significant issue in the WAAM of alloys based on aluminium [56]. Because of the creation of gas pores during the deposition of even single-layer beads, typical CMT-WAAM is also inappropriate for material fabrication. Cong et al. [57] evaluated various CMT modes, including normal CMT, CMT-P (CMT pulse), CMT-ADV (CMT advanced), and CMT-PADV (CMT pulse advanced), while examining the impact of the arcing mode in CMT-WAAM on the creation of porosity in AlCu6 alloy. Pores larger than 100 μm were created in substantial quantities using conventional CMT. CMT-P had significantly superior performance, with no pores larger than 100 μm. In this regard, CMT-ADV and CMT-PADV performed very well, as no pores larger than 50 μm and 10 μm, respectively, were

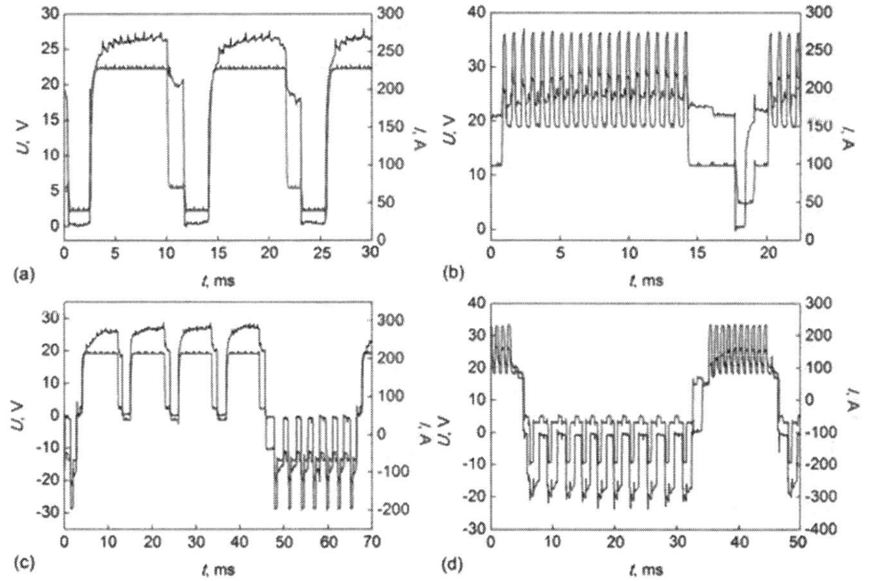

FIGURE 5.10 Graphical illustration of arc voltage and current for (a) CMT, (b) CMT-P, (c) CMT-ADV, and (d) CMT-PADV methods [57]. (Reprinted by permission of Springer.)

discovered. Lesser heat input, fine and equiaxed grain formation, less dilution, and oxide cleaning of the wire are all credited with the removal of porosity. Figure 5.10 displays the arc voltage and current waveforms of several CMT-WAAM modes on keeping the wire-feed rate constant. Similar findings of fewer pores in an Al alloy were reported by Cong et al. [58] utilising various arc modes of CMT-WAAM.

5.4.3 Step-Over Deposition

The ability of welding-oriented techniques to overlap gives them a distinct edge. These processes can result in the formation of a supportless molten pool under the surface tension produced through the already cemented layer. The fundamental difficulty with welding-based technologies is maintaining the transferring material stability and weld formation during deposition. Using a torch at an angle [59] and giving an offset on the welding torch [60] can help in solving these issues. To comprehend the material transfer behaviour of the step-over deposition technique generated slant structures by CMT-WAAM, Yan et al. [61] invented slanted short-circuiting transfer (SSCT). Figure 5.11 displays the diagrammatic design of the experimental rig made up of welding, sensing, and photography devices. By adjusting the step-over distance and SSCT, a slanted structure was created, eliminating the need for any support structures. This step-over deposition capability has to be further researched since it could be essential for developing useful 3-D components using the CMT-based WAAM approach.

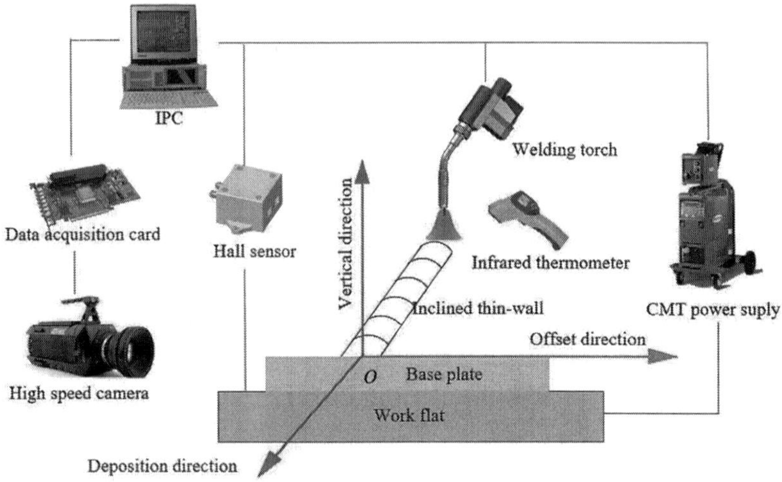

FIGURE 5.11 Diagrammatic presentation of methodology and rig of step-over deposition [61]. (Reprinted by permission of Elsevier.)

5.5 CONCLUSION

Based on the excellent potential and capabilities of the CMT-WAAM technique, it has captured the attention of industries and researchers to use it as a mainstream additive manufacturing technique. In this chapter, an overview of the CMT-WAAM process with emphasis on their recent technical developments is presented. This chapter starts with the introduction of wire arc additive manufacturing technologies and highlights the need for CMT-based WAAM by addressing the advantages of CMT-WAAM over other contemporary WAAM methods. CMT was detailed with its mechanism, and the materials fabricated with CMT-WAAM, including steels, titanium, aluminum, and nickel-based alloys, were discussed. Various recent advances in this field, including neat immersion active cooling, CMT pulse advanced, and step-over deposition, were discussed in the last portion of this chapter.

REFERENCES

[1] B. Wu, Z. Pan, D. Ding, D. Cuiuri, H. Li, J. Xu, J. Norrish, A review of the wire arc additive manufacturing of metals: Properties, defects and quality improvement, *J. Manuf. Process.* 35 (2018) 127–139. https://doi.org/10.1016/j.jmapro.2018.08.001

[2] B. Ralph, Method of making decorative articles. United States, US1533300A. Filed Nov. 12, 1920, Granted April 14, 1925.

[3] C.R. Cunningham, J.M. Flynn, A. Shokrani, V. Dhokia, S.T. Newman, Invited review article: Strategies and processes for high quality wire arc additive manufacturing, *Addit. Manuf.* 22 (2018) 672–686. https://doi.org/10.1016/j.addma.2018.06.020

[4] S. Pattanayak, S.K. Sahoo, Gas metal arc welding based additive manufacturing—A review, *CIRP J. Manuf. Sci. Technol.* 33 (2021) 398–442. https://doi.org/10.1016/j.cirpj.2021.04.010

[5] P. Dickens, M. Pridham, R. Cobb, I. Gibson, G. Dixon, Rapid prototyping using 3-D welding, in DTIC Document, (1992). https://doi.org/hdl.handle.net/2152/64409
[6] J.D. Spencer, P.M. Dickens, C.M. Wykes, Rapid prototyping of metal parts by three-dimensional welding, *Proc. Inst. Mech. Eng. Part B J. Eng. Manuf.* 212 (1998) 175–182. https://doi.org/10.1243/0954405981515590
[7] B. Tomar, S. Shiva, T. Nath, A review on wire arc additive manufacturing: Processing parameters, defects, quality improvement and recent advances, *Mater. Today Commun.* 31 (2022) 103739. https://doi.org/10.1016/j.mtcomm.2022.103739
[8] B. Tomar, Cold metal transfer-based wire arc additive manufacturing, *J. Brazilian Soc. Mech. Sci. Eng.* 2 (2023). https://doi.org/10.1007/s40430-023-04084-2
[9] K. Furukawa, New CMT arc welding process – welding of steel to aluminium dissimilar metals and welding of super-thin aluminium sheets, *Weld. Int.* 20 (2006) 440–445. https://doi.org/10.1533/wint.2006.3598
[10] S. Kou, *Welding Metallurgy*, John Wiley & Sons, Inc., Hoboken, NJ, 2002. https://doi.org/10.1002/0471434027
[11] S.J. Bless, Drop transfer in short-circuit welding, *J. Phys. D. Appl. Phys.* 7 (1974) 306. https://doi.org/10.1088/0022-3727/7/4/306
[12] C.G. Pickin, K. Young, Evaluation of cold metal transfer (CMT) process for welding aluminium alloy, *Sci. Technol. Weld. Join.* 11 (2006) 583–585. https://doi.org/10.1179/174329306X120886
[13] J. Feng, H. Zhang, P. He, The CMT short-circuiting metal transfer process and its use in thin aluminium sheets welding, *Mater. Des.* 30 (2009) 1850–1852. https://doi.org/10.1016/j.matdes.2008.07.015
[14] Z. Sun, Y. Lv, B. Xu, Y. Liu, J. Lin, K. Wang, Investigation of droplet transfer behaviours in cold metal transfer (CMT) process on welding Ti-6Al-4V alloy, *Int. J. Adv. Manuf. Technol.* 80 (2015) 2007–2014. https://doi.org/10.1007/s00170-015-7197-9
[15] B. Mezrag, F. Deschaux-Beaume, M. Benachour, Control of mass and heat transfer for steel/aluminium joining using cold metal transfer process, *Sci. Technol. Weld. Join.* 20 (2015) 189–198. https://doi.org/10.1179/1362171814Y.0000000271
[16] S. Selvi, A. Vishvaksenan, E. Rajasekar, Cold metal transfer (CMT) technology – An overview, *Def. Technol.* 14 (2018) 28–44. https://doi.org/10.1016/j.dt.2017.08.002
[17] O. Panchenko, D. Kurushkin, I. Mushnikov, A. Khismatullin, A. Popovich, A high-performance WAAM process for Al–Mg–Mn using controlled short-circuiting metal transfer at increased wire feed rate and increased travel speed, *Mater. Des.* 195 (2020) 109040. https://doi.org/10.1016/j.matdes.2020.109040
[18] P. Wang, H. Zhang, H. Zhu, Q. Li, M. Feng, Wire-arc additive manufacturing of AZ31 magnesium alloy fabricated by cold metal transfer heat source: Processing, microstructure, and mechanical behavior, *J. Mater. Process. Technol.* 288 (2021) 116895. https://doi.org/10.1016/j.jmatprotec.2020.116895
[19] K. Pal, S.K. Pal, Effect of pulse parameters on weld quality in pulsed gas metal arc welding: A review, *J. Mater. Eng. Perform.* 20 (2011) 918–931. https://doi.org/10.1007/s11665-010-9717-y
[20] H. Stinson, R. Ward, J. Quinn, C. McGarrigle, Comparison of properties and bead geometry in mig and cmt single layer samples for waam applications, *Metals (Basel)* 11 (2021). https://doi.org/10.3390/met11101530
[21] P. Wang, S. Hu, J. Shen, Y. Liang, Characterization the contribution and limitation of the characteristic processing parameters in cold metal transfer deposition of an Al alloy, *J. Mater. Process. Technol.* 245 (2017) 122–133. https://doi.org/10.1016/j.jmatprotec.2017.02.019

[22] E.M. Ryan, T.J. Sabin, J.F. Watts, M.J. Whiting, The influence of build parameters and wire batch on porosity of wire and arc additive manufactured aluminium alloy 2319, *J. Mater. Process. Technol.* 262 (2018) 577–584. https://doi.org/10.1016/j.jmatprotec.2018.07.030

[23] J. Sun, J. Hensel, M. Köhler, K. Dilger, Residual stress in wire and arc additively manufactured aluminum components, *J. Manuf. Process.* 65 (2021) 97–111. https://doi.org/10.1016/j.jmapro.2021.02.021

[24] X. Fang, L. Zhang, J. Yang, H. Bai, L. Zhao, K. Huang, B. Lu, Effect of characteristic substrate parameters on the deposition geometry of CMT additive manufactured Al-6.3%Cu alloy, *Appl. Therm. Eng.* 162 (2019) 114302. https://doi.org/10.1016/j.applthermaleng.2019.114302

[25] M. Dong, Y. Zhao, Q. Li, Y. Fei, T. Zhao, F. Wang, A. Wu, Microstructure evolution and mechanical property anisotropy of wire and arc-additive-manufactured wall structure using ER2319 welding wires, *J. Mater. Eng. Perform.* 30 (2021) 258–268. https://doi.org/10.1007/s11665-020-05336-1

[26] J. Chen, H. Wei, X. Zhang, Y. Peng, J. Kong, K. Wang, Flow behavior and microstructure evolution during dynamic deformation of 316 L stainless steel fabricated by wire and arc additive manufacturing, *Mater. Des.* 198 (2021) 109325. https://doi.org/10.1016/j.matdes.2020.109325

[27] J. Chen, H. Wei, K. Bao, X. Zhang, Y. Cao, Y. Peng, J. Kong, K. Wang, Dynamic mechanical properties of 316L stainless steel fabricated by an additive manufacturing process, *J. Mater. Res. Technol.* 11 (2021) 170–179. https://doi.org/10.1016/j.jmrt.2020.12.097

[28] Y. Ali, P. Henckell, J. Hildebrand, J. Reimann, J.P. Bergmann, S. Barnikol-Oettler, Wire arc additive manufacturing of hot work tool steel with CMT process, *J. Mater. Process. Technol.* 269 (2019) 109–116. https://doi.org/10.1016/j.jmatprotec.2019.01.034

[29] J. Ge, T. Ma, Y. Chen, T. Jin, H. Fu, R. Xiao, Y. Lei, J. Lin, Wire-arc additive manufacturing H13 part: 3D pore distribution, microstructural evolution, and mechanical performances, *J. Alloys Compd.* 783 (2019) 145–155. https://doi.org/10.1016/j.jallcom.2018.12.274

[30] B. Tomar, S. Shiva, Microstructure evolution in steel/copper graded deposition prepared using wire arc additive manufacturing, *Mater. Lett.* 328 (2022) 133217. https://doi.org/10.1016/j.matlet.2022.133217

[31] Z. Wang, A.M. Beese, Effect of chemistry on martensitic phase transformation kinetics and resulting properties of additively manufactured stainless steel, *Acta Mater.* 131 (2017) 410–422. https://doi.org/10.1016/j.actamat.2017.04.022

[32] J. Ge, J. Lin, Y. Lei, H. Fu, Location-related thermal history, microstructure, and mechanical properties of arc additively manufactured 2Cr13 steel using cold metal transfer welding, *Mater. Sci. Eng. A.* 715 (2018) 144–153. https://doi.org/10.1016/j.msea.2017.12.076

[33] N. Rodriguez, L. Vázquez, I. Huarte, E. Arruti, I. Tabernero, P. Alvarez, Wire and arc additive manufacturing: A comparison between CMT and TopTIG processes applied to stainless steel, *Weld. World.* 62 (2018) 1083–1096. https://doi.org/10.1007/s40194-018-0606-6

[34] L. Sun, F. Jiang, R. Huang, D. Yuan, C. Guo, J. Wang, Anisotropic mechanical properties and deformation behavior of low-carbon high-strength steel component fabricated by wire and arc additive manufacturing, *Mater. Sci. Eng. A.* 787 (2020) 139514. https://doi.org/10.1016/j.msea.2020.139514

[35] P.P. Nikam, D. Arun, K.D. Ramkumar, N. Sivashanmugam, Microstructure characterization and tensile properties of CMT-based wire plus arc additive manufactured ER2594, *Mater. Charact.* 169 (2020) 110671. https://doi.org/10.1016/j.matchar.2020.110671

[36] Y. Wang, X. Chen, S.V. Konovalov, Additive manufacturing based on welding arc: A low-cost method, *J. Surf. Investig. X-Ray, Synchrotron Neutron Tech.* 11 (2017) 1317–1328. https://doi.org/10.1134/S1027451017060210

[37] B. Baufeld, Mechanical Properties of INCONEL 718 Parts Manufactured by Shaped Metal Deposition (SMD), *J. Mater. Eng. Perform.* 21 (2012) 1416–1421. https://doi.org/10.1007/s11665-011-0009-y

[38] R.M. Kindermann, M.J. Roy, R. Morana, P.B. Prangnell, Process response of Inconel 718 to wire + arc additive manufacturing with cold metal transfer, *Mater. Des.* 195 (2020) 109031. https://doi.org/10.1016/j.matdes.2020.109031

[39] Y. Wang, X. Chen, Q. Shen, C. Su, Y. Zhang, S. Jayalakshmi, R.A. Singh, Effect of magnetic Field on the microstructure and mechanical properties of inconel 625 superalloy fabricated by wire arc additive manufacturing, *J. Manuf. Process.* 64 (2021) 10 19. https://doi.org/10.1016/j.jmapro.2021.01.008

[40] D. Van, G.P. Dinda, J. Park, J. Mazumder, S.H. Lee, Enhancing hardness of Inconel 718 deposits using the aging effects of cold metal transfer-based additive manufacturing, *Mater. Sci. Eng. A.* 776 (2020) 139005. https://doi.org/10.1016/j.msea.2020.139005

[41] Y. Tian, J. Shen, S. Hu, J. Gou, Y. Cui, Effects of cold metal transfer mode on the reaction layer of wire and arc additive-manufactured Ti-6Al-4V/Al-6.25Cu dissimilar alloys, *J. Mater. Sci. Technol.* 74 (2021) 35–45. https://doi.org/10.1016/j.jmst.2020.09.014

[42] N. Xu, J. Shen, S. Hu, Y. Tian, J. Bi, Bimetallic structure of Ti6Al4V and Al6.21Cu fabricated by cold metal transfer additive manufacturing via Nb interlayer added by TIG, *Mater. Lett.* 302 (2021) 130397. https://doi.org/10.1016/j.matlet.2021.130397

[43] A. Antonysamy, Microstructure, Texture and Mechanical Property Evolution during Additive Manufacturing of Ti6Al4V Alloy for Aerospace Applications, Ph.D. Thesis, University of Manchester, Manchester, UK (2012).

[44] S. Zhou, H. Xie, J. Ni, G. Yang, L. Qin, X. Guo, Metal transfer behavior during CMT-based wire arc additive manufacturing of Ti-6Al-4V alloy, *J. Manuf. Process.* 82 (2022) 159–173. https://doi.org/10.1016/j.jmapro.2022.07.063

[45] A. Caballero, J. Ding, Y. Bandari, S. Williams, Oxidation of Ti-6Al-4V during wire and arc additive manufacture, 3D print. *Addit. Manuf.* 6 (2019) 91–98. https://doi.org/10.1089/3dp.2017.0144

[46] B. Wu, D. Ding, Z. Pan, D. Cuiuri, H. Li, J. Han, Z. Fei, Effects of heat accumulation on the arc characteristics and metal transfer behavior in Wire Arc Additive Manufacturing of Ti6Al4V, *J. Mater. Process. Technol.* 250 (2017) 304–312. https://doi.org/10.1016/j.jmatprotec.2017.07.037

[47] ASTM F1472-20a, Standard Specification for Wrought Ti-6Al-4V alloy for Surgical Implant Applications (UNS R56400), ASTM International, West Conshohocken, PA (2020). https://doi.org/10.1520/F1472-20A

[48] ASTM F1108-21, Standard Specification for Ti-6Al-4V alloy Castings for Surgical Implants (UNS R56406), ASTM International, West Conshohocken, PA (2021). https://doi.org/10.1520/F1108-21

[49] L. Vazquez, M.N. Rodriguez, I. Rodriguez, P. Alvarez, Influence of post-deposition heat treatments on the microstructure and tensile properties of ti-6al 4v parts manufactured by cmt-waam, *Metals (Basel)* 11 (2021). https://doi.org/10.3390/met11081161

[50] G. Pardal, F. Martina, S. Williams, Laser stabilization of GMAW additive manufacturing of Ti-6Al-4V components, *J. Mater. Process. Technol.* 272 (2019) 1–8. https://doi.org/10.1016/j.jmatprotec.2019.04.036

[51] B. Zheng, Y. Zhou, J.E. Smugeresky, J.M. Schoenung, E.J. Lavernia, Thermal behavior and microstructural evolution during laser deposition with laser-engineered net shaping: Part I. Numerical calculations, *Metall. Mater. Trans. A Phys. Metall. Mater. Sci.* 39 (2008) 2228–2236. https://doi.org/10.1007/s11661-008-9557-7

[52] A. Vahedi Nemani, M. Ghaffari, S. Salahi, J. Lunde, A. Nasiri, Effect of interpass temperature on the formation of retained austenite in a wire arc additive manufactured ER420 martensitic stainless steel, *Mater. Chem. Phys.* 266 (2021). https://doi.org/10.1016/j.matchemphys.2021.124555

[53] L.J. da Silva, H.N. Ferraresi, D.B. Araújo, R.P. Reis, A. Scotti, Effect of thermal management approaches on geometry and productivity of thin-walled structures of ER 5356 built by wire + arc additive manufacturing, *Coatings* 11 (2021) 1141. https://doi.org/10.3390/coatings11091141

[54] L.J. da Silva, D.M. Souza, D.B. de Araújo, R.P. Reis, A. Scotti, Concept and validation of an active cooling technique to mitigate heat accumulation in WAAM, *Int. J. Adv. Manuf. Technol.* 107 (2020) 2513–2523. https://doi.org/10.1007/s00170-020-05201-4

[55] F.M. Scotti, F.R. Teixeira, L.J. da Silva, D.B. de Araújo, R.P. Reis, A. Scotti, Thermal management in WAAM through the CMT Advanced process and an active cooling technique, *J. Manuf. Process.* 57 (2020) 23–35. https://doi.org/10.1016/j.jmapro.2020.06.007

[56] B. Cong, R. Ouyang, B. Qi, J. Ding, Influence of cold metal transfer process and its heat input on weld bead geometry and porosity of aluminum-copper alloy welds, *Rare Met. Mater. Eng.* 45 (2016) 606–611. https://doi.org/10.1016/S1875-5372(16)30080-7

[57] B. Cong, J. Ding, S. Williams, Effect of arc mode in cold metal transfer process on porosity of additively manufactured Al-6.3%Cu alloy, *Int. J. Adv. Manuf. Technol.* 76 (2015) 1593–1606. https://doi.org/10.1007/s00170-014-6346-x

[58] B. Cong, Z. Qi, B. Qi, H. Sun, G. Zhao, J. Ding, A comparative study of additively manufactured thin wall and block structure with Al-6.3%Cu alloy using cold metal transfer process, *Appl. Sci.* 7 (2017). https://doi.org/10.3390/APP7030275

[59] P. Kazanas, P. Deherkar, P. Almeida, H. Lockett, S. Williams, Fabrication of geometrical features using wire and arc additive manufacture, *Proc. Inst. Mech. Eng. Part B J. Eng. Manuf.* 226 (2012) 1042–1051. https://doi.org/10.1177/0954405412437126

[60] J.S. Panchagnula, S. Simhambhatla, Manufacture of complex thin-walled metallic objects using weld-deposition based additive manufacturing, *Robot. Comput. Integr. Manuf.* 49 (2018) 194–203. https://doi.org/10.1016/j.rcim.2017.06.003

[61] Z. Yan, Y. Zhao, F. Jiang, S. Chen, F. Li, W. Cheng, X. Ma, Metal transfer behaviour of CMT-based step-over deposition in fabricating slant features, *J. Manuf. Process.* 71 (2021) 147–155. https://doi.org/10.1016/j.jmapro.2021.09.027

6 Metal Alloys and Beyond
Analysing the Horizon of WAAM Materials

Amrit Raj Paul
Royal Melbourne Institute of Technology, Melbourne, Australia
Academy of Scientific and Innovative Research, Ghaziabad, India

Manidipto Mukherjee
CSIR-Central Mechanical Engineering Research Institute, Durgapur, India

6.1 INTRODUCTION

The WAAM process utilises an electric arc to melt metal wire and build 3D objects layer by layer. It commonly employs a wide range of materials like aluminium, titanium, stainless steel, and nickel alloys, and recently, refractory metals and non-metals have been introduced. These materials are chosen for their strength, durability, and resistance to harsh environments. The sustainable development of WAAM requires knowledge of these materials to produce complex metal parts for industries such as aerospace, automotive, and biomedical applications. The chapter explores the evolving use of materials in WAAM, including the shift towards powder or hybrid forms and their impact on mechanical and functional properties.

6.1.1 Wire Arc Additive Manufacturing (WAAM)

An example of a directed energy deposition (DED) additive manufacturing system is WAAM, which uses an electric arc as a high-energy source and wire as the raw material. The gas metal arc welding (GMAW) or gas tungsten arc welding (GTAW) processes are somewhat comparable to this technology. In GMAW, an electric arc is created between the electrode and the work piece, which generates a tremendous amount of heat and causes the material to melt and deposit at the welding zone along with the electrode. The metal is fed to the arc in the form of wire in GTAW, where the electrode is non-consumable. Additionally, an inert gas atmosphere is offered to prevent hydrogen embrittlement and component oxidation [1]. An electric arc is produced by running an electric current through a tungsten electrode and used in the WAAM process. The electrode is placed close to the substrate or existing part's surface, and as current passes through it, a high-temperature arc is created that melts the

DOI: 10.1201/9781003363415-6

wire feedstock. The desired object is then constructed by layering the molten metal. The main distinction between WAAM and arc welding, despite their many similarities, is the material's intended use. Arc welding is used when the base plate (substrate) is in bulk, and the deposited material (weld bead) is in the minority. WAAM is used when the base plate (substrate) is in the minority, and the deposited material (weld bead) is in the majority. Figure 6.1 provides a schematic diagram to illustrate the differences between the two processes.

Conventional parts like a welding power source, an arc-generating torch, and an automated wire feeding system frequently make up the hardware needed for WAAM. A computerised robotic system is able to control the tool path and wire feed rate with extreme precision. Depending on the complexity of the shape being created, the robotic arm can move in a range of directions and incorporates both the tool and the wire feeding system. Figure 6.2 depicts a typical WAAM setup (available at WAAM3DP, CSIR-CMERI).

The technology of additive manufacturing has advanced significantly over the last two decades and is predicted to do so for at least another century. The advantages of WAAM over other additive manufacturing processes are numerous. In contrast to techniques like DED, which only achieve a deposition rate of 14%–20%, WAAM achieves the highest deposition rate, ranging from 90% to 98%. Higher yields, less material waste, and shorter lead times for the finished product are the results of this. In their research on high-quality metal additive manufacturing, Paul Colegrove and

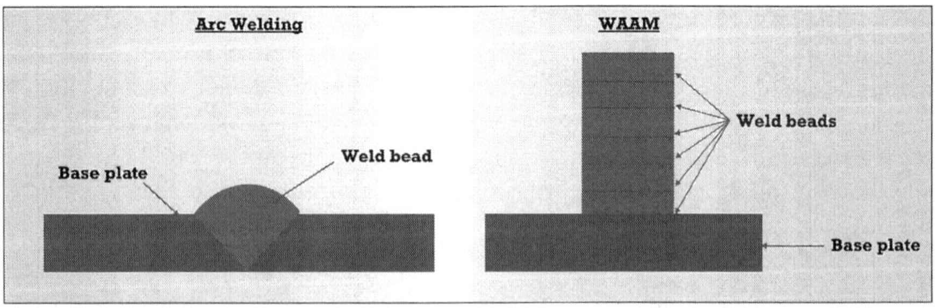

FIGURE 6.1 Schematic differentiation between arc welding and WAAM.

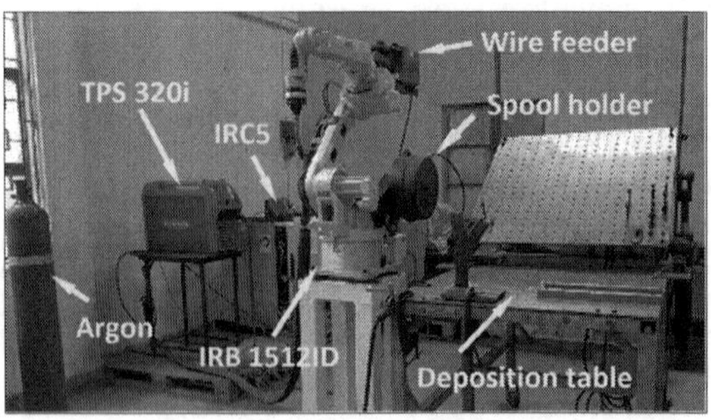

FIGURE 6.2 Typical WAAM arrangement. (Available at CSIR-CMERI.)

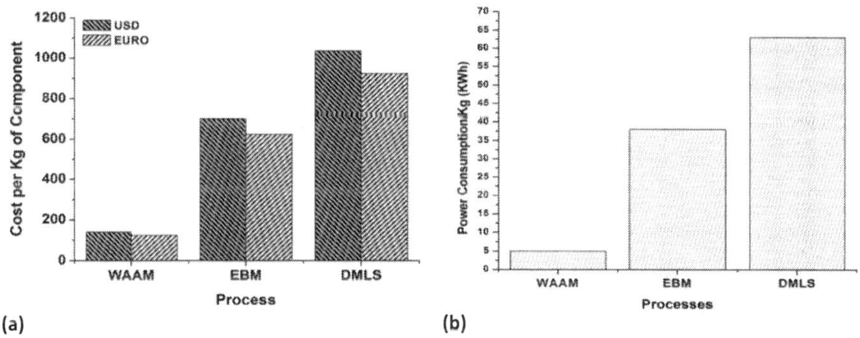

FIGURE 6.3 (a) Cost comparison [3] and (b) power consumption comparison between different AM techniques [3].

Professor Stewart Williams of Cranefield University demonstrated the superior deposition rates of WAAM [2]. Moreover, WAAM requires relatively low initial capital costs since it employs a traditional metal inert gas (MIG) welding system. In comparison, techniques such as LASER or electron beam systems incur higher costs. Additionally, other additive manufacturing techniques primarily use costly metal powders as raw materials, while WAAM uses raw materials in the form of wires that are less expensive. In a comparative analysis of WAAM, electrical beam melting (EBM), and direct metal laser sintering (DMLS) presented by Bekker et al. [3], as shown in Figure 6.3a, the manufacturing cost per kilogram of the component by WAAM was approximately 23% of the cost incurred by EBM and approximately 14% of the cost incurred by DMLS. Thirdly, WAAM is energy-efficient as it requires approximately 85% less energy than powder-based deposition technology. Although the total energy consumption between wire-based and powder-based additive manufacturing (AM) is almost similar, the energy required for the deposition process in WAAM is significantly less compared to other additive manufacturing processes. A study conducted by Bekker et al. [3] indicated that the energy consumption in the deposition process of WAAM is highly efficient, as shown in Figure 6.3b. Furthermore, WAAM gives designers more design freedom because it can work with a variety of ferrous and non-ferrous metals, giving them a wide range of materials to choose from depending on the mechanical properties they need. The final product has more design freedom because the component's product geometry is solely dependent on the tool path. In general, WAAM as an AM technology offers a number of benefits, including a high deposition rate and a low capital cost, making it a viable option for producing large structures with high quality and efficiency.

6.1.2 Deposition Techniques in WAAM

WAAM is a highly versatile and efficient AM process that involves the deposition of molten material onto a substrate using an electric arc. There are three primary deposition techniques used in WAAM: conventional arc-based deposition technique, cold metal transfer (CMT)-deposition technique, and plasma transferred arc (PTA)–based deposition technique.

6.1.2.1 Conventional Arc-Based Deposition Technique

Two well-known and common deposition techniques that are frequently used in the WAAM process are MIG and TIG (tungsten inert gas). A consumable wire electrode is fed continuously into the welding pool when using the MIG technique. The wire is heated by an electrical arc between it and the workpiece, and the molten metal that results is used to deposit metal onto a substrate layer by layer, constructing the final part [4]. Simultaneously, TIG-based WAAM is a process that uses a non-consumable tungsten electrode to produce the electrical arc. The electrode is held in a torch, and a separate filler wire is fed into the welding pool [4]. The resulting molten metal is used to deposit metal onto a substrate layer by layer, assembling the finished part. Shielding gases like He, Ar, N_2, CO_2, etc., are used to shield the molten metal pool from atmospheric gas dissolution while deposition is occurring. Figure 6.4 depicts a schematic representation of the MIG and TIG-based deposition techniques used for the WAAM process. Standard arc-based WAAM, if compared to other metal AM techniques, has a number of benefits, including high deposition rates and adaptability. It is a versatile process that can be applied to a wide range of metals. Its primary drawback is that it results in a rough surface finish that might need additional post-processing.

In summary, MIG and TIG-based WAAM processes are two popular methods for depositing metal layer by layer onto a substrate. Both processes have advantages and disadvantages, and the choice between them will depend on the specific requirements of the part being fabricated. MIG-based WAAM is typically faster and can be used with a wide range of metals but may require additional post-processing to achieve a smooth surface finish [5]. TIG-based WAAM produces a comparatively smoother surface finish but may be slower and more limited in the range of metals that can be used [6].

6.1.2.2 CMT-Based Deposition Technique

The CMT process is another type of wire-based WAAM technique that has been gaining popularity in recent years. It is a hybrid welding process that combines the heating characteristics of both MIG and TIG welding. This process uses a specially designed wire feed mechanism that controls the wire feed rate and arc current

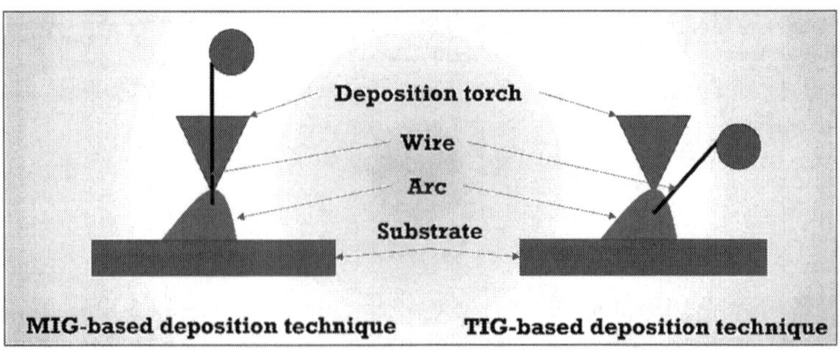

FIGURE 6.4 MIG and TIG-based deposition techniques used in the WAAM process.

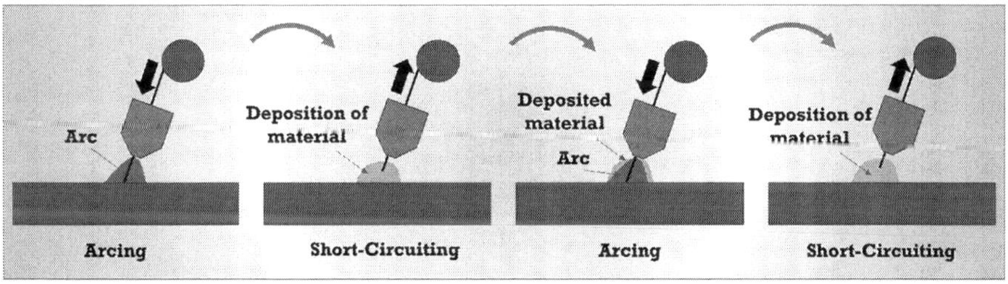

FIGURE 6.5 CMT deposition technique for WAAM process.

simultaneously, allowing for greater precision in the deposition process. In the CMT process, a high-frequency pulse is applied to the wire as it is fed through the welding torch. This pulse causes the wire to "short-circuit" with the workpiece, creating a small droplet of molten metal that is then transferred to the workpiece. The pulse is then turned off, allowing the arc to extinguish and the droplet to solidify [7]. Figure 6.5 illustrates the CMT deposition technique process schematically. Rapid repetition of this procedure causes a continuous deposition of material. Controlling heat input is one of the CMT process' key benefits because it lessens the possibility of distortion or warping in the substrate or deposited structure [8]. This makes it especially beneficial for the deposition of materials like Al, Cu, Ti, and other alloys that are sensitive to atmospheric oxygen. Because the controlled heat input helps to prevent the formation of brittle intermetallic compounds at dissimilar interfaces, the CMT process can also be used to create transitional multi-material structures [9]. Another benefit of the CMT process is its ability to reduce spatter and fumes. The controlled wire feed and current make it possible to achieve a stable arc, which reduces the amount of spatter and fumes generated during the welding process. This makes the CMT process a safer and more environmentally friendly option for metal deposition [7, 8]. Overall, metal deposition in WAAM applications using the CMT process is a promising option. It is an adaptable and powerful option for a variety of applications due to its capacity to control heat input, lessen spatter and fumes, and weld metals that are not the same alloy.

6.1.2.3 PTA-Based Deposition Technique

An electric arc is used in the PTA-based WAAM process, a metal AM technology, to deposit material layer by layer. The arc welding process, which has been used for many years in industry to join and repair metal parts, is a variation of the technology in question. A wire made of the material to be deposited is first introduced into the system as part of the PTA-based WAAM process. Continuous wire feeding is used, and the system is controlled to keep the wire's distance from the substrate constant. The substrate is typically a metal plate, and it is mounted on a movable platform that allows for precise control of the deposition process [10]. Once the wire is positioned, the PTA system is activated. The PTA system consists of a plasma torch that generates a plasma arc, which is directed onto the wire electrode. The plasma arc is created

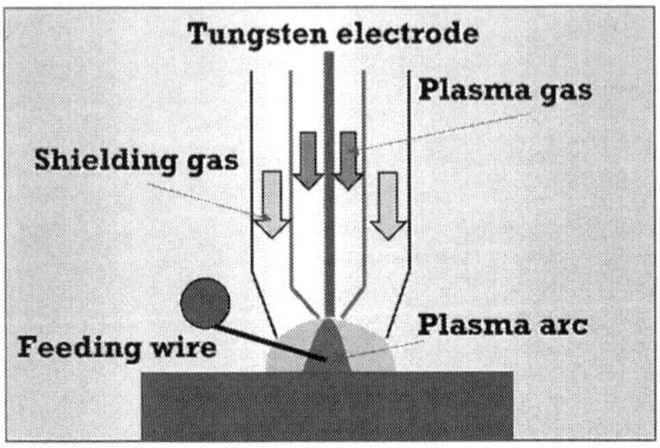

FIGURE 6.6 PTA-based deposition technique for WAAM.

by passing an inert gas, typically argon, through a high-frequency electric arc. This generates temperatures of up to 20,000°C [10, 11], which melts the fed wire. Then, a controlled deposition of the molten metal from the wire onto the substrate occurs. The movable platform moves the substrate along the horizontal and vertical axes to give it the desired shape. The wire's feed rate and the movable platform's speed regulate the rate of deposition. Figure 6.6 provides a schematic illustration of the PTA-based deposition procedure for WAAM application.

High-strength materials like nickel-based alloys, titanium alloys, and stainless steels can be deposited using the PTA-based WAAM process with great success [11]. Due to their high melting points and high atmospheric reactivity, these materials are challenging to process using other AM methods. In comparison to other AM technologies, the WAAM process based on PTA has a number of benefits. The ability to create outsized and intricate geometries is one benefit [8]. The deposition process is managed by a mobile platform, allowing the system to produce parts with complex geometries and high levels of complexity. This is particularly helpful for creating unique parts or fixing broken parts. Producing parts with excellent metallurgical properties is another benefit of the PTA-based WAAM process [12]. The precise control of the process parameters enables the production of high-quality parts, while the high-energy plasma arc aids in the reduction of porosity and other defects [10]. The final part's strength and toughness are increased by the very fine grain structure of the deposited metal [12]. The PTA-based WAAM process has a number of industrial applications. Making high-strength components for use in aerospace and automobile applications is one such application. The procedure can also be used to fix broken parts, like engine parts or turbine blades. Additionally, custom parts like moulds or dies can be produced using the PTA-based WAAM process.

6.1.3 WAAM Materials

A wide variety of materials can be deposited using the flexible and economical WAAM process. Metals, ceramics, and composites are just a few of the materials to which this technology has been applied. The choice of material is determined by the demands of the application, such as mechanical characteristics, thermal characteristics, and resistance to corrosion.

Metals are the most commonly used materials in WAAM processes. These include:

Steel: The most common metal used in WAAM processes is steel. It is a reasonably priced and incredibly adaptable material that has a wide range of uses, including structural elements, machine parts, and tools. Different kinds of steel, such as low-carbon, medium-carbon, and high-carbon steel, have been deposited using WAAM processes [12–14]. By modifying the wire's chemical composition, the welding settings, and the heat treatment procedure, the mechanical properties of the deposited steel can be specifically tailored.

Aluminium: Aluminium is a corrosion-resistant, lightweight metal that is frequently used in automotive and aerospace applications. It is highly recyclable and exhibits excellent thermal and electrical conductivity. Different types of aluminium, including pure aluminium, aluminium alloys, and composite materials, have been deposited using WAAM processes [15, 16]. Parts with a low weight-to-strength ratio, good thermal conductivity, and better corrosion resistance can be made using the deposited aluminium.

Titanium: High-strength and lightweight titanium is a metal that is frequently employed in aerospace and medical applications. It is biocompatible and has superior resistance to corrosion, making it the perfect material for use in implant materials. Different types of titanium, including pure titanium, titanium alloys, and composite materials, have been deposited using WAAM processes [17, 18]. Parts with a low weight-to-strength ratio, good biocompatibility, and exceptional corrosion resistance can be made using the deposited titanium.

Nickel-based alloys: Due to their excellent mechanical characteristics, resistance to corrosion, and heat resistance, nickel-based alloys are a popular choice for WAAM and are suitable for a variety of applications. They frequently appear in high-temperature components like gas turbines and jet engines [19–21].

Copper: Copper is an excellent conductor of heat and electricity and is commonly used in electrical and electronic applications. It is also highly corrosion-resistant and has good mechanical properties. WAAM processes have been used to deposit various types of copper, including pure copper and copper alloys [22]. The deposited copper can be used to produce parts with excellent electrical conductivity, thermal conductivity, and corrosion resistance.

Magnesium: Magnesium is a thin, lightweight metal that is frequently used in aerospace and automotive products. It is easily recyclable and has a great strength-to-weight ratio. AZ31, AZ61, and ZK60 are just a few of the different magnesium alloys that have been deposited using WAAM processes [23, 24]. Parts with a low weight-to-strength ratio, strong resistance to corrosion, and excellent damping capacity can be made using the deposited magnesium.

In conclusion, a variety of metals, such as steels, aluminium, titanium, nickel-based alloys, copper, and magnesium, can be processed using the WAAM method.

Parts with specific mechanical characteristics, such as a low weight-to-strength ratio, high resistance to corrosion, and strong electrical conductivity, can be made using the deposited metal. The application requirements, such as the mechanical characteristics, thermal characteristics, and resistance to corrosion, determine the metal to be used. By enabling the production of complex parts with shorter lead times and lower costs, WAAM methods have the potential to completely transform the manufacturing sector.

6.1.4 Need to Study WAAM Materials

WAAM is a cost-effective way of producing large metal components with design flexibility. The quality of WAAM parts depends on material selection, shielding gas composition, and processing parameters. Studying WAAM materials can help optimise these factors and reduce production costs. Understanding the behaviour of different materials during the WAAM process can help optimise part design and produce high-quality parts. The following is the point-wise description to justify the need to study WAAM materials.

Process optimisation: Understanding how materials behave during the WAAM process can help optimise the process's variables, improving product quality while cutting down on production time and expense.

Material selection: Parts made from a range of materials, such as metals, alloys, and composites, can be produced using WAAM. The best material for a given application can be chosen by studying these materials' characteristics.

Performance evaluation: The unique microstructure produced by the AM process may cause the properties of WAAM materials to differ from those of materials produced conventionally. Examining these characteristics can help determine the effectiveness of WAAM parts and guarantee that they adhere to the necessary requirements.

Quality control: WAAM materials may have defects, such as porosity, cracks, or distortion. Understanding the causes of these defects and developing methods to control or eliminate them is crucial for ensuring the quality of WAAM products.

In general, studying the WAAM materials is crucial for developing the technology and making sure that it is successfully applied in a variety of industries, such as aerospace, automotive, and medical.

6.2 TYPES OF MATERIALS USED IN WAAM

WAAM has the versatility to use most metals that can be drawn into wire, including those commonly used for arc welding. A percentage breakup of different typical materials used for the WAAM process is shown in Figure 6.7 (data collected from Web of Science™). It can be clearly seen that the most researched materials in the WAAM process are stainless steel, Titanium and alloys, Aluminium and alloys, and nickel and alloys, followed in the same order. However, in recent years, some wires have been specifically developed for the WAAM process, such as nitinol,

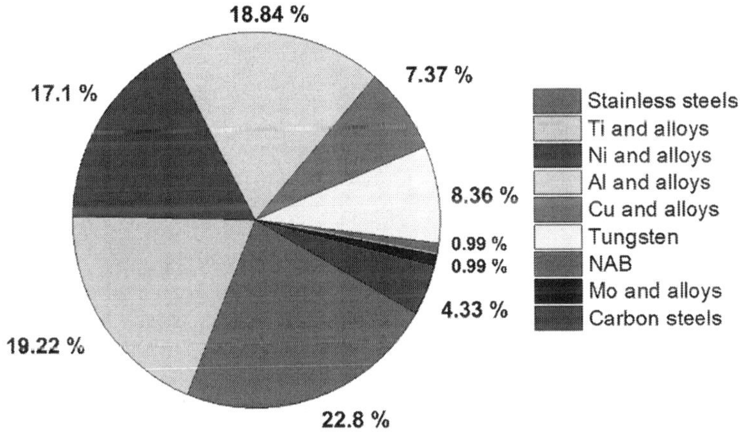

FIGURE 6.7 Breakup of materials used in the WAAM process. (Data collected from Web of Science™.)

tungsten, stellite, CuNi, Invar, and others. These materials have been implemented into WAAM processes for their unique metallurgical and mechanical properties. While locally available MIG-based welding wires can be used, there are global players who specialise in providing wires designed specifically for the WAAM process. Companies like Sandvik, Voestalpine Böhler Welding, Lincoln Electric, WAAM3D, and ESAB offer a wide range of wires that meet the requirements of WAAM. This section provides a comprehensive overview of the different materials used to advance the WAAM process.

6.2.1 Steels

Steel is an extremely versatile material for WAAM, with a specific choice of steel depending on the unique requirements of each application. By studying the properties and behaviour of different steels during the WAAM process, it is possible to optimise process parameters and improve the overall quality of the final product. Over time, various types of steels have been tested and proven to have superior metallurgical and mechanical properties when used in the WAAM process. Typical types of steels that have been used in the context of WAAM are as follows:

- *Carbon steels*: Carbon steels are one of the materials which have been mostly used in the WAAM process. In fact, the superior drawability of low-carbon steels over other carbon steels makes it the big contender to be used as a WAAM material. Steel with a carbon content of less than 0.25% is ideal for producing ductile and tough parts through WAAM. This type of steel is commonly designated by the nomenclature ERXS-Y, where E stands for electrode, R for round, X denotes the minimum tensile strength in PSI, S refers to solid, and Y denotes the level of deoxidisers and silicon in the matrix. Table 6.1 (AWS 5.18 2005; AWS 5.28 2020) provides a

TABLE 6.1
Chemical Composition of Different Carbon Steels

Nomenclature	Chemical Composition (wt. %)													Reference Standard
	C	*Mn*	*Si*	*P*	*S*	*Ni*	*Cr*	*Mo*	*V*	*Cu*	*Ti*	*Zr*	*Al*	
ER70S-2	0.07	0.90–1.40	0.40–0.70	0.025	0.035	0.15	0.15	0.15	0.03	0.50	0.05–0.15	0.02–0.12	0.05–0.15	AWS 5.18
ER70S-3	0.06–0.15	0.90–1.40	0.45–0.75	0.025	0.035	0.15	0.15	0.15	0.03	0.50	—	—	—	AWS 5.18
ER70S-4	0.06–0.15	1.00–1.50	0.65–0.85	0.025	0.035	0.15	0.15	0.15	0.03	0.50	—	—	—	AWS 5.18
ER70S-6	0.06–0.15	1.40–1.85	0.80–1.15	0.025	0.035	0.15	0.15	0.15	0.03	0.50	—	—	—	AWS 5.18
ER70S-7	0.07–0.15	1.50–2.00	0.50–0.80	0.025	0.035	0.15	0.15	0.15	0.03	0.50	—	—	—	AWS 5.18
ER70S-8	0.02–0.10	1.40–1.90	0.55–1.10	0.025	0.035	0.15	0.15	0.15	0.03	0.50	—	—	—	AWS 5.18
ER70S-A1	0.12	1.30	0.30–0.70	0.025	0.025	0.20	—	0.40–0.65	—	0.35	—	—	—	AWS 5.28
ER80S-B2	0.07–0.12	0.40–0.70	0.40–0.70	0.025	0.025	0.20	1.20–1.50	0.40–0.65	—	0.35	—	—	—	AWS 5.28
ER80S-B3L	0.05	0.40–0.70	0.40–0.70	0.025	0.025	0.20	2.30–2.70	0.90–1.20	—	0.35	—	—	—	AWS 5.28
ER80S-Ni1	0.12	1.25	0.40–0.80	0.025	0.025	0.80–1.10	0.15	0.35	0.05	0.35	—	—	—	AWS 5.28
ER90S-B3	0.07–0.12	0.40–0.70	0.40–0.70	0.025	0.025	0.20	2.30–2.70	0.90–1.20	—	0.35	—	—	—	AWS 5.28
ER90S-D2	0.07–0.12	1.60–2.10	0.50–0.80	0.025	0.025	0.15	—	0.40–0.60	—	0.50	—	—	—	AWS 5.28
ER100S-1	0.08	1.25–1.80	0.20–0.55	0.010	0.010	1.40–2.10	0.30	0.25–0.55	0.05	0.25	0.10	0.10	0.10	AWS 5.28
ER110S-1	0.09	1.40–1.80	0.20–0.55	0.010	0.010	1.90–2.60	0.50	0.25–0.55	0.04	0.25	0.10	0.10	0.10	AWS 5.28
ER120S-1	0.10	1.40–0.80	0.25–0.60	0.010	0.010	2.00–2.80	0.60	0.30–0.65	0.03	0.25	0.10	0.10	0.10	AWS 5.28
ER120S-G	0.09	1.9	0.80	0.05	0.05	2.15	0.35	0.55	0.04	0.08	0.012	0.10	0.13	AWS 5.28

summary of the chemical composition of commonly used low-carbon steel wires. Among the low-carbon steel variants, ER70S-6 has received significant research attention for WAAM applications. According to Clarivate's (Web of Science™) database, 45 research articles have been published on low-carbon steel, of which 35 pertain to ER70S. Figure 6.8a provides a comprehensive breakdown of low-carbon steel wire types that are suitable for use in the WAAM process. Low-carbon steel typically exhibits a ferritic-pearlitic structure, which is illustrated in Figure 6.8b. In this figure, the whitish grains represent pearlite, while the blackish grain boundaries are ferrites. The pearlite adds toughness and ductility, while the ferrites add strength and hardness to the substance. Depending on the desired mechanical characteristics, the appropriate material series must be chosen for the final deposited structure. The ER70 series is one of the most popular options because of its low alloy and carbon content. In particular, the ER70S-6 variant is especially appealing for WAAM research because it is easily accessible, reasonably priced, compatible with a variety of materials, and has a long history of welding research.

Experts frequently choose low-carbon steel for structural applications because it is inexpensive, readily available, and has adequate mechanical

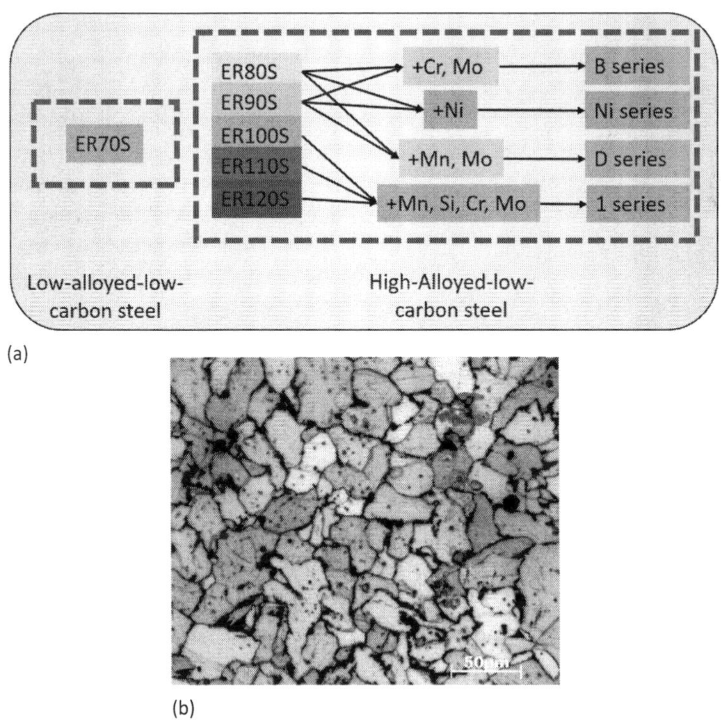

FIGURE 6.8 (a) Low-carbon steel materials for WAAM and (b) ferritic-pearlitic microstructure of low-carbon steel.

properties. For some applications, the material's limited lifespan due to susceptibility to corrosion is a significant disadvantage. Low-carbon steel's strength-to-weight ratio is also frequently an issue. Stainless steel provides a solution in these circumstances.

- *Stainless steel*: Due to its superior mechanical characteristics, corrosion-resistant properties, and durability, stainless steel is one of the materials that WAAM uses the most frequently. Different types of stainless steel are available for use in the production of WAAMed components, as shown in Table 6.2. Due to their outstanding corrosion resistance and high ductility, austenitic stainless steels—among the various types of stainless steel that are currently available—are typically the type of stainless steel that is used most frequently in WAAM. In summary, there are four primary types of stainless steel: austenitic stainless steel (300 series), ferritic and martensitic stainless steel (400 series), duplex (austenitic+ferritic) stainless steel (2000 series), and precipitation-hardening (PH) martensitic stainless steel (600 series) [25]. The types of stainless steel can be determined using Scheffler's diagram, which relies on the nickel equivalence (Ni_{eq}) and chromium equivalence (Cr_{eq}) formulas [26]. Figure 6.9a shows a representative Scheffler diagram, while Equations 6.1 and 6.2 [26] provide the formulas for Ni_{eq} and Cr_{eq}. By increasing the respective equivalence, the material's corresponding property can be developed since Ni_{eq} is known as an austenitic stabiliser, and Cr_{eq} is known as a ferrite stabiliser. Table 6.3 [27, 28] provides a property-based analysis. Increasing the amount of austenite in the metal matrix can enhance the material's corrosion resistance and weldability during deposition. However, this comes at a cost, as it can decrease the material's strength compared to the other grades of stainless steel [29]. A typical microstructure of austenitic stainless steel and duplex stainless steel is shown in Figure 6.9b and c, respectively. The white portion in the micrograph is austenite, whereas the black portion is ferrite. Thus, it is possible to significantly alter the mechanical characteristics of stainless steel by adjusting the austenite and ferrite content through a customised chemical composition. For instance, ferritic stainless steel's corrosion resistance is improved by adding chromium, while austenitic stainless steel's ductility and toughness are improved by adding nickel. Controlling the heat treatment procedure can also have an impact on the mechanical characteristics of stainless steel, including its ductility, strength, and hardness. The mechanical properties of stainless steel can be customised to meet particular requirements for a variety of applications by enhancing the chemical composition and heat treatment process.

$$Ni_{eq} = Ni + 30C + 0.5Mn \tag{6.1}$$

$$Cr_{eq} = Cr + Mo + 1.5Si + 0.5Nb \tag{6.2}$$

TABLE 6.2
Chemical Compositions of Different Stainless Steels (Composition % w/w)

Nomenclature	C	Mn	Si	Cr	Mo	Ni	Nb+Ta	N	S	P	Cu	Fe	Reference Standard
ER307	0.07	6.60	0.90	18.70	—	8.90	—	—	0.007	0.015	—	Bal	ISO 14343
ER308H	0.05	1.65	0.46	20.45	—	9.85	—	—	0.005	0.016	—	Bal	AWS A5.9
ER308L	0.019	1.72	0.46	20.80	—	10.10	—	—	0.003	0.013	—	Bal	AWS A5.9
ER308Lsi	0.016	1.65	0.85	20.65	—	10.00	—	—	0.008	0.016	—	Bal	AWS A5.9
ER309L	0.021	1.75	0.51	23.40	—	12.70	—	—	0.006	0.014	—	Bal	AWS A5.9
ER309Lsi	0.019	1.85	0.84	23.50	—	12.95	—	—	0.004	0.012	—	Bal	AWS A5.9
ER310	0.11	1.90	0.40	38.10	—	20.95	—	—	0.003	0.012	—	Bal	AWS A5.9
ER312	0.11	1.64	0.44	29.60	—	8.90	—	—	0.012	0.017	—	Bal	AWS A5.9
ER316L	0.016	1.87	0.48	19.32	2.25	13.20	—	—	0.010	0.019	—	Bal	AWS A5.9
ER316Lsi	0.022	1.80	0.85	19.25	2.45	12.60	—	—	0.004	0.013	—	Bal	AWS A5.9
ER317L	0.017	1.66	0.44	19.40	3.25	13.85	—	—	0.006	0.012	—	Bal	AWS A5.9
ER320LR	0.025	1.60	0.05	19.60	2.50	34.10	0.25	—	0.001	0.007	3.40	Bal	AWS A5.9
ER330	0.23	1.95	0.42	15.95	—	35.20	—	—	0.005	0.014	—	Bal	AWS A5.9
ER347	0.04	1.65	0.52	19.90	—	9.75	0.72	—	0.005	0.014	—	Bal	AWS A5.9
ER385	0.019	2.05	0.35	20.50	4.60	25.10		—	0.015	0.014	1.60	Bal	AWS A5.9
ER410NiMo	0.05	0.62	0.48	11.50	—	—	0.35	—	0.016	0.018	—	Bal	AWS A5.9
ER410	0.11	0.45	0.39	12.50	—	—	—	—	0.010	0.014	—	Bal	AWS A5.9
ER420	0.29	0.45	0.35	13.20	—	—	—	—	0.008	0.014	—	Bal	AWS A5.9
ER430	0.07	0.44	0.36	16.50	—	—	—	—	0.010	0.014	—	Bal	AWS A5.9
ER430LCb	0.01	0.51	0.40	18.20	—	—	0.40	—	0.008	0.020	—	Bal	AWS A5.9
ER630	0.03	0.54	0.43	16.49	0.20	4.78	0.22	—	0.021	0.017	3.60	Bal	AWS A5.9
ER2209	0.016	1.40	0.45	22.40	3.20	8.50	—	0.18	0.017	0.014	—	Bal	AWS A5.9
ER2594	0.07	1.00	1.00	25.50	3.95	9.25	—	0.25	0.010	0.025	0.50	Bal	AWS A5.9

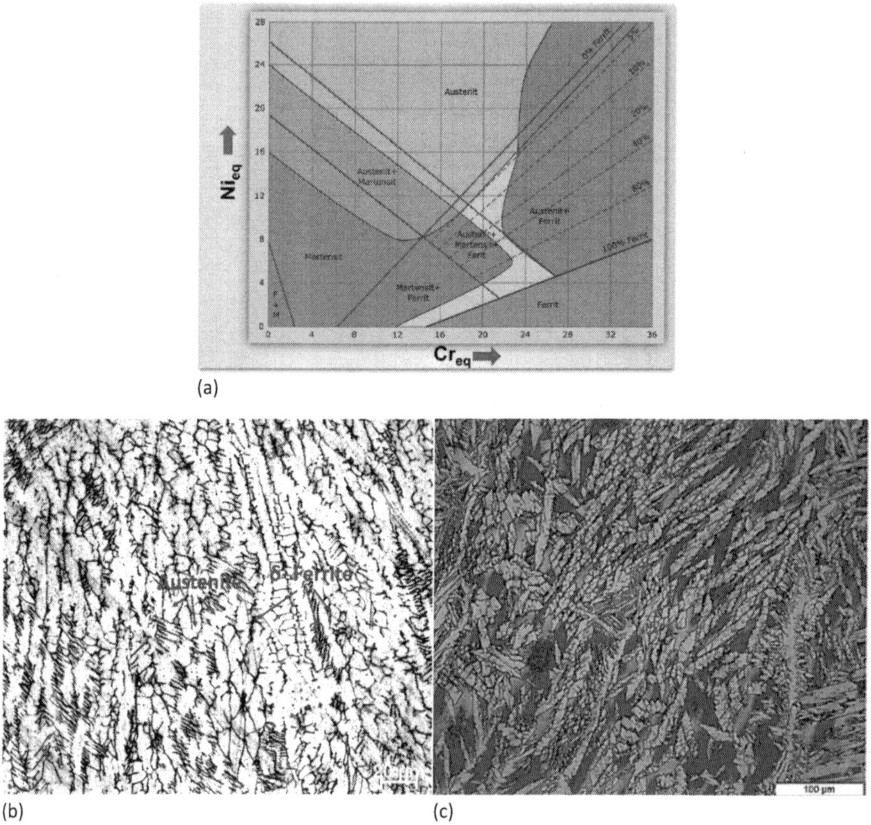

FIGURE 6.9 (a) Schaeffler diagram, (b) microstructure of SS316L deposited through WAAM, and (c) microstructure of SS2594 deposited through WAAM ([a] adapted from Zhang et al. [26], under CC-By 4.0 license and [b] adapted from Sales et al. [29], under CC-By 4.0 license)

TABLE 6.3
Properties Associated with Different Grades of Stainless Steel (SS)

SS Grade/ Properties	Weldability	Ductility	Refractoriness	Corrosion Resistance	Common Grades
Austenitic	High	High	High	High	304, 308, 316
Ferritic	Low	Moderate	High	Moderate	409, 430, 444
Martensitic	Low	Low	Low	Low	410, 416, 420
Duplex	High	Moderate	Low	Low	2209, 2304, 2594
PH	Low	Moderate	Low	Low	630

The WAAM process, in conclusion, is a promising method for creating stainless steel parts with distinctive properties that can satisfy a variety of industrial requirements. The anisotropy, residual stress, and distortion that may be present in WAAMed parts can be reduced by using in situ rolling techniques. By optimising the WAAM and heat treatment parameters, the microstructure and mechanical characteristics of parts made of stainless steel that have undergone WAAM can be controlled. Furthermore, WAAMed parts may have exceptional fatigue characteristics, making them appropriate for applications involving cyclic loading, such as those found in aircraft and automobile parts. Parts made of WAAMed are ideal for harsh environments like chemical or marine applications because their corrosion resistance can be customised to meet particular needs. The versatility of stainless-steel parts made using this technique will increase with further application range expansion and WAAM process optimisation.

6.2.2 Aluminium and Alloys

Aluminium is a corrosion-resistant, lightweight metal that has gained popularity in a wide range of industries, such as construction, automotive, and aerospace [30]. WAAM is one of the AM technologies that has made it possible to fabricate intricate aluminium parts with superior mechanical qualities [8]. Its outstanding thermal and electrical conductivity, low weight, high specific strength, remarkable formability, and high specific strength all contribute to its many positive qualities. Its density is only 2.7 g/cm^3 [31], which is one-third that of steel, and its face-centred cubic (FCC) crystal structure, which has a low susceptibility to dislocation slip, makes it highly workable. The creation of a thick, self-healing oxide coating explains its excellent corrosion resistance in oxidising situations. Aluminium's strength, weldability, and formability can be tailored by alloying components for particular uses. Copper, manganese, silicon, magnesium, and zinc are examples of common solutes that can be added [32]. Cast alloys and wrought alloys are the two categories into which aluminium alloys fall. Cast alloys are rarely used as wire consumables for arc deposition because they are substantially alloyed for increased castability and to obtain reasonable strength. Thus, in accordance with the availability of wiring raw materials, the focus has been limited to wrought alloys. Figure 6.10 illustrates the division of wrought alloys into seven series depending on the main alloying constituents.

The 1000-series aluminium is commercially pure with less than 1% of additional elements present, which are frequently contaminants from the initial production process. This class of alloys is rarely used for WAAM due to its softness. Nonetheless, this series is appropriate for cladding applications due to its superior resistance to corrosion [32]. The 2000-series features high-strength but weakly corrosion-resistant aluminium-copper alloys. Atomic clusters (Guinier-Preston zones) and plate-shaped Al_2Cu precipitates created following heat treatment give these alloys their mechanical qualities. The initial purpose of the 2319 alloy wire consumables was to link 2219 plates and forgings. The fatigue-resistant 2024 alloy, together with the alloys 2319 and 2219, are suitable for WAAM. Manganese serves as the primary alloying element in the alloys of the 3000-series. These are universal alloys for beverage cans with excellent formability and middle strength levels. WAAM never treated these alloys, and they were rarely utilised as filler in welding operations. The intermediate

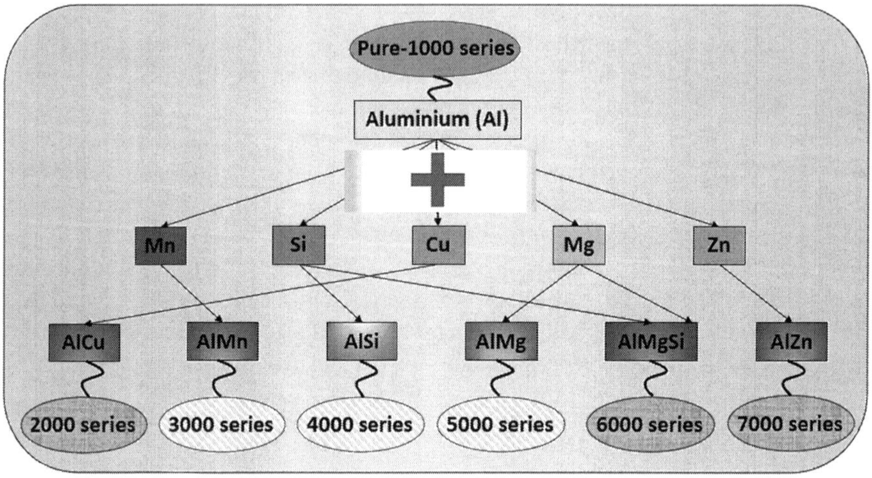

FIGURE 6.10 Breakup of aluminium alloys.

strength values of the 4000-series aluminium alloys result from silicon strengthening in a solid solution [33]. Because of their high flowability and minimal thermal shrinkage, Al-Si alloys are primarily utilised for casting. The most popular welding and brazing techniques are used for wrought applications. Due to its outstanding weldability with a variety of aluminium alloys, the 4043 alloy is the backbone of arc welding and is utilised in a number of industries. As a result, Al4043 is extensively researched for WAAM [32]. Magnesium serves as the primary alloying element in 5000-series aluminium alloys, while trace amounts of manganese are added to improve the dispersion. These alloys are frequently employed in shipbuilding because the magnesium concentration boosts their corrosion resistance in maritime settings. Al-Mg alloys experience severe strain hardening as a result of the Portevin-Le Chatelier effect [34]. In the following years, with commercial feedstock, alloys from the 5000-series, such as 5087, 5183, and 5356, repeatedly exhibited their good characteristics in WAAM materials [30]. Al-Mg-Si alloys of the 6000-series can be heat treated for improved strength by precipitating semi-coherent Mg_2Si (β' and β''-precipitates). Many uses exist for these alloys, particularly for extruded profiles. Nevertheless, processing 6000-series alloys via AM is challenging due to their high susceptibility to breaking during solidification [35]. Commercial feedstock is currently in short supply for both the WAAM and welding processes. The only Al-Mg-Si alloy that is readily available and appropriate for use as a weld filler, to the best of the authors' knowledge, is 6063 [36]. The impressive mechanical qualities of 7000-series aluminium alloys, which include zinc as the main alloying element and balanced proportions of magnesium and copper, are attained through the precipitation of $MgZn_2$ and Mg (Zn, Cu, Al)$_2$ η-phases after artificial heat treatment. Military aviation applications commonly use the 7068 alloys, the strongest commercially available aluminium alloy [36]. Nevertheless, the 7000-series alloys needed for WAAM are not currently offered commercially.

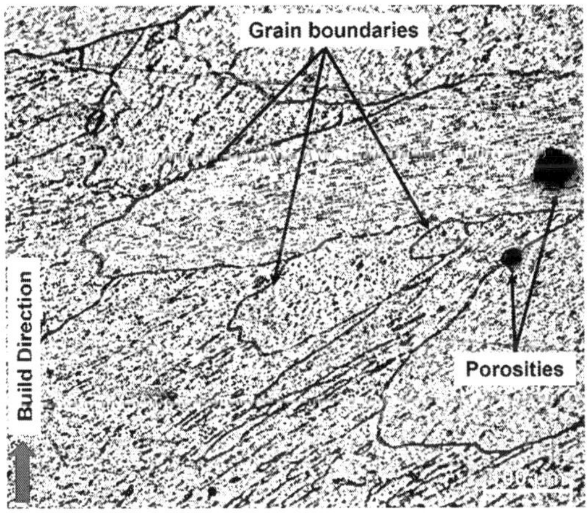

FIGURE 6.11 Grain morphology and porosity evolution of Al4043 deposited through WAAM.

It has been demonstrated that using aluminium via WAAM results in products with mechanical qualities that are on par with or occasionally superior to those of wrought or cast goods. Notwithstanding these developments, there are still challenges with grain shape, porosity, solidification cracking, and residual stress in WAAM-processed aluminium alloys, which contribute to the mechanical and microstructural variability of these alloys [37]. Figure 6.11 depicts the development of porosity and grain morphology. However, by carefully regulating variables like current and voltage, choosing appropriate process modes like arc pulsation and CMT, and making process advancements such as interlayer rolling, researchers have been successful in overcoming these problems [38]. Faster cooling rates and lower heat input have also resulted in the development of equiaxed structures with reduced porosity and enhanced solidification cracking resistance [37]. These techniques advance the use of WAAM in the automotive and aerospace industries by considerably enhancing microstructural uniformity and mechanical characteristics.

6.2.3 Copper and Alloys

A class of metallic materials known as copper alloys is made of copper with one or more other elements. Due to its great thermal and electrical conductivity and high ductility and malleability, copper is a valuable material in a variety of applications. It is a well-liked option for outdoor and marine applications because it is also corrosion-resistant. Due to its excellent conductivity, low resistance, and low thermal expansion, copper is frequently used in electrical wire and other electrical components. Moreover, it has excellent heat conductivity, which makes it perfect for use in heat sinks and other thermal management applications. Moreover, copper is frequently utilised in water supply and plumbing systems. Copper pipes and fittings are a well-liked option for both residential and commercial plumbing systems since they are strong and corrosion-resistant. Because of its superior heat transfer capabilities, copper is also utilised in heating systems like radiators.

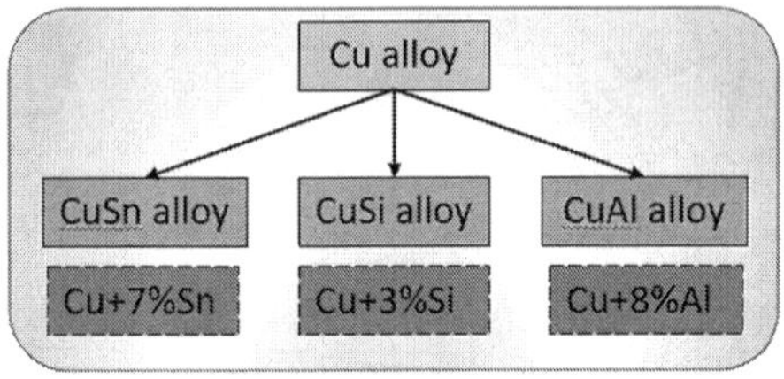

FIGURE 6.12 Different Cu alloys used for the WAAM process.

With respect to the WAAM process, primarily, three types of copper alloys have been used so far, including CuSn alloys, CuSi alloys, and CuAl alloys (see Figure 6.12). The drawability and varying post-deposition strength of these alloys make their wires easily accessible for WAAM applications. CuSn, CuSi, and CuAl alloys have mechanical strengths of about 220 MPa, 350 MPa, and 420 MPa, respectively, when connected. As a result of its superior qualities compared to the other two alloys, CuSi has been the most frequently used of these three alloys for the WAAM process up to this point. CuSn alloys may not be used because of their poor strength, while CuAl alloys may not be used because they are prone to the formation of CuAl intermetallics, which can make the structure more brittle. The WAAM technique enables the economical exploration and development of copper alloys while lowering production costs. Compared to conventional manufacturing methods, WAAM has a number of benefits, such as faster production, less material waste, and more design flexibility. Furthermore, the WAAM technique has tremendous research potential for copper alloys. By adopting this method, it is possible to produce novel and distinctive copper alloys with improved performance and performance characteristics that are not currently possible using conventional manufacturing methods. We can lower the cost of copper components for a variety of technical applications by creating these alloys.

In summary, copper alloys are a broad class of metallic materials with special qualities that make them indispensable in a variety of industries. With lowering production costs, the WAAM technique provides the ability to investigate and develop these alloys cheaply. Also, the WAAM technique offers tremendous research potential for copper alloys, which will result in novel and intriguing applications in a variety of industries. We can lower the cost of copper components for many engineering applications by creating new copper alloys with improved qualities and performance attributes.

6.2.4 Nickel and Alloys

The utilisation of a variety of materials, including nickel alloys, is one of the main benefits of WAAM. Because of their outstanding high-temperature characteristics, corrosion resistance, and mechanical strength, nickel alloys are especially well suited for WAAM [39]. In a variety of industrial applications, such as the aerospace, chemical, oil and gas, and marine industries, the use of nickel alloys in WAAM has attracted considerable attention. Many engineering applications frequently include nickel alloys, some of which are as follows:

- Inconel: A family of nickel-based superalloys called Inconel is renowned for its exceptional resistance to corrosion and high temperatures. In high-temperature applications, including gas turbine parts, heat exchangers, and nuclear reactors, Inconel alloys are frequently employed. High concentrations of nickel, chromium, and iron are frequently found in Inconel alloys, along with other substances, including molybdenum, cobalt, and titanium [40, 41]. Inconel alloys, in comparison to other superalloys, have good resistance to oxidation and corrosion in high-temperature environments, making them appropriate for usage in challenging circumstances.
- Rene: Rene is a family of nickel-based superalloys that are renowned for their corrosion resistance as well as their thermal strength. Rene alloys are typically employed in high-stress environments like gas turbine engines and aerospace parts. Rene alloys often contain substances that give them their distinct qualities, including nickel, cobalt, chromium, molybdenum, tungsten, and aluminium [42]. Rene alloys are more suited for use in applications where high-temperature strength is essential than Inconel because they exhibit stronger strength and creep resistance at high temperatures.
- Waspaloy: In high-temperature and high-stress applications, including gas turbine engines, aircraft engines, and nuclear reactors, Waspaloy, a superalloy based on nickel, is employed. High concentrations of nickel, chromium, and molybdenum, together with lesser levels of titanium and aluminium, can be found in Waspaloy [43]. Waspaloy is excellent for use in applications where high-temperature strength and durability are essential because it has stronger strength and better resistance to fatigue and creep at high temperatures than Inconel [43].
- Hastelloy: A series of nickel-based superalloys called Hastelloy is renowned for its exceptional strength at high temperatures and resistance to corrosion. Hastelloy alloys are frequently employed in applications for power production, pollution control, and chemical processing, among others. Nickel, molybdenum, chromium, iron, and other elements, including tungsten, cobalt, and copper, are frequently found in Hastelloy alloys [44, 45]. Hastelloy alloys are superior to Inconel in corrosion resistance in corrosive conditions, making them appropriate for use in chemical processing and other corrosive applications.
- Hynes: Hynes is a superalloy with a nickel base that is typically employed in high-temperature applications such as gas turbine engines, power generation, and aerospace parts. High concentrations of nickel, cobalt, chromium, and tungsten are found in Hynes, along with lesser levels of other elements like titanium and aluminium [46]. Hynes is a good choice for use in situations where high-temperature strength and durability are important since it has stronger strength and better resistance to elevated-temperature oxidation and corrosion than Inconel.

- Incoloy: A class of nickel-based superalloys known as Incoloy is employed in a number of corrosive and high-temperature processes, such as oil and gas exploration, chemical processing, and power production. High concentrations of nickel, chromium, and iron are frequently found in Incoloy alloys, along with additional substances, including molybdenum, copper, and titanium [47, 48]. Incoloy alloys outperform Inconel in elevated-temperature environments in terms of resistance to corrosion, making them appropriate for usage in demanding situations where these requirements are essential.

These are just a few examples of the nickel alloys that are commonly used in WAAM. The specific alloy used will depend on the application and the properties required for the finished product. The available wired electrodes of Ni alloys for the WAAM process are tabulated in Table 6.4. Out of these available alloys, only superalloy categories such as Inconel 625 and Inconel 718 have gained more popularity because of their superior high-temperature strength, much relevant to powerplant and aerospace applications. Moreover, amongst the superalloy category, the Inconel 625 and Inconel 718 are much studied as these two are readily available and comparatively cheaper than the other superalloys.

Nickel, chromium, molybdenum, and niobium are the primary constituents in Inconel 625, but they also contain trace amounts of iron, cobalt, and carbon. Together, these components improve the material's mechanical characteristics and ability to withstand high temperatures and corrosive conditions. With a tensile strength of 773 MPa and yield strength of 646 MPa at room temperature, Inconel 625 has excellent mechanical properties [49]. Moreover, it has exceptional fatigue resistance and can survive high-stress and high-temperature conditions. Extreme conditions, such as those encountered in the chemical processing and marine engineering industries, do not readily corrode or oxidise Inconel 625 [50]. It can be used in seawater and other harsh conditions since it is resistant to pitting and crevice corrosion. Inconel 625 is appropriate for usage in high-temperature situations, including heat exchangers, boiler hardware, and exhaust systems because it is resistant to corrosion and oxidation at high temperatures [50]. Inconel 625 is a solid-solution-strengthened alloy, which means that the crystal lattice structure of the alloy is strengthened by the addition of nickel and other constituents. The alloy's microstructure consists of a combination of austenite and gamma double prime (γ'') (Ni_3Nb) phases [49], as shown in Figure 6.13a. On the other hand, Inconel 718 is composed of nickel, chromium, molybdenum, niobium, iron, and other trace elements, including titanium and aluminium. Inconel 718 is a precipitation-hardened alloy [53], which means that it is strengthened by the formation of a secondary phase through heat treatment. The alloy's microstructure consists of a combination of gamma prime (γ') (Ni_3Al/Ti) and gamma double prime (γ'') (Ni_3Nb) phases [53]. An example of a gamma double prime phase is shown in Figure 6.13b. High tensile and yield strength, strong fatigue strength, and great ductility are just a few of Inconel 718's remarkable mechanical characteristics. With a tensile strength of 775 MPa and yield strength of 412 MPa at room temperature, Inconel 718 has exceptional mechanical qualities [53].

TABLE 6.4
Available Nickel-Based Alloys for WAAM Process (Composition % w/w)

AWS Nomenclature	Common Name	C	Mn	Fe	Cu	Si	P	S	Ti	Al	Mo	Ta+Nb	Co	Cr	Ni	Reference
ERNi-1	Ni2061	0.15	1.0	1.0	0.25	0.75	0.03	0.015	2.0–3.5	1.5	—	—	—	—	Bal.	AWS5.14
ERNiCu-7	Monel 400	0.15	4.0	2.5	Bal.	1.25	0.02	0.015	1.5–3.0	1.25	—	—	—	—	62.0–69.0	AWS5.14
ERNiCu-8	Monel 500	0.25	1.5	2.0	Bal.	1.0	0.03	0.015	0.25–1.0	2.0–4.0	—	—	—	—	63.0–70.0	AWS5.14
ERNiCr-3	Ni6082	0.10	2.5–3.5	3.0	0.50	0.50	0.03	0.015	0.75	—	—	2.0–3.0	—	18.0–22.0	Bal.	AWS5.14
ERNiCr-4	Ni6072	0.01–0.10	0.2	0.50	0.50	0.20	0.02	0.015	0.3–1.0	—	—	—	—	42.0–46.0	Bal.	AWS5.14
ERNiCrFe-5	Ni6062	0.08	1.0	10.0	0.50	0.35	0.03	0.015	—	—	—	1.5–3.0	—	14.0–17.0	Bal.	AWS5.14
ERNiCrFe-7	Ni6052	0.04	1.0	7.0–11.0	0.30	0.50	0.02	0.015	1.0	1.10	0.50	0.10	—	28.0–31.0	Bal.	AWS5.14
ERNiCrFe-12	Ni6025	0.15–0.25	0.50	8.0–11.0	0.1	0.5	0.02	0.01	0.1–0.2	1.8–2.4	—	—	1.0	24.0–26.0	Bal.	AWS5.14
ERNiCoCrSi-1	Hynes 160	0.05–0.12	1.0	21.0–25.0	0.3	2.5–3.0	0.02	0.01	—	0.3	—	—	1.0	26.0–29.0	Bal.	AWS5.14
ERNiCrFeAl-1	Inconel 693	0.15	1.0	2.5–6.0	0.5	0.5	0.03	0.01	1.0	2.0–4.0	—	0.5–2.5	—	27.0–31.0	Bal.	AWS5.14
ERNiFeCr-1	Incoloy 65	0.05	1.0	22.0	0.3	0.5	0.03	0.03	0.6–1.2	0.20	2.5–3.5	—	—	19.5–23.5	Bal.	AWS5.14
ERNiFeCr-2	Inconel 718	0.08	0.35	Bal.	0.3	0.35	0.015	0.015	0.65–1.15	0.2–0.8	2.8–3.3	4.75–5.5	—	17.0–21.0	50.0–55.0	AWS5.14
ERNiMo-1	-	0.08	1.0	4.0–7.0	0.5	1.0	0.025	0.03	—	—	26.0–30.0	—	2.5	1.0	Bal.	AWS5.14
ERNiMo-3	Hastelloy W	0.12	1.0	4.0–7.0	0.5	1.0	0.04	0.03	—	—	23.0–26.0	—	—	4.0–6.0	Bal.	AWS5.14
ERNiMo-10	Hastelloy B-3	0.01	3.0	1.0–3.0	0.2	0.1	0.03	0.01	0.2	0.5	27.0–32.0	0.2	3.0	1.0–3.0	Bal.	AWS5.14
ERNiCrMo-2	Hastelloy X	0.05–0.15	1.0	17.0–20.0	0.5	1.0	0.04	0.03	—	—	8.0–10.0	—	0.5–2.5	20.5–23.0	Bal.	AWS5.14
ERNiCrMo-3	Inconel 625	0.1	0.5	5.0	0.5	0.5	0.02	0.015	0.4	0.4	8.0–10.0	3.15–4.15	—	20.0–23.0	Bal.	AWS5.14
ERNiCrMo-17	Hastelloy C-2000	0.01	0.5	3.0	1.3–1.9	0.08	0.025	0.01	—	0.5	15.0–17.0	—	2.0	22.0–24.0	Bal.	AWS5.14
ERNiCrCoMo-1	Inconel 617	0.05–0.15	1.0	3.0	0.5	1.0	0.03	0.015	0.6	0.8–1.5	8.0–10.0	—	10.0–15.0	20.0–24.0	Bal.	AWS5.14

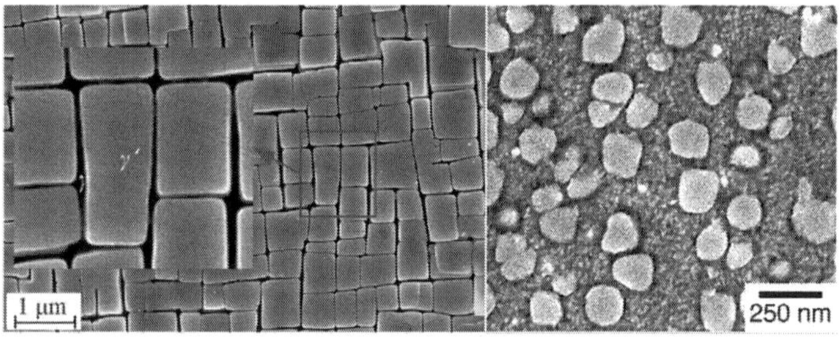

(a) (b)

FIGURE 6.13 Microstructure of (a) gamma prime (γ′) and (b) gamma double prime (γ″). ([a] adapted from Feng et al. [51], under CC-By 4.0. license and [b] adapted from Mignanelli et al. [52], under CC-By 4.0 license.)

Moreover, it has exceptional fatigue resistance and can survive high-stress and high-temperature conditions. It is suited for usage in high-stress and high-temperature situations because of these characteristics. Extreme environments, including those found in the aerospace, petrol turbine, and nuclear power industries, do not affect Inconel 718's ability to resist corrosion or oxidation. In addition, it shares Inconel 625's resistance to pitting and crevice corrosion, making it appropriate for usage in harsh environments like seawater. In addition to the common Inconel 718, Allegheny Technologies Incorporated (ATI) has developed ATI 718PLUS, a superalloy based on nickel that has undergone PH to improve its weldability and high-temperature capabilities. This alloy has been specifically created to maintain the processing properties of Inconel 718 while exhibiting the excellent heat stability and temperature capabilities of Waspaloy. Although maintaining the majority of the advantages of its base alloy, Inconel 718, it offers thermal stability up to 55°C higher than Inconel 718. It is believed that major changes to the composition of Inconel 718, which led to the creation of a more stable strengthening phase (γ′), are to blame for this alloy's extraordinary improvement in thermal stability. The eta (η) phase has been identified as the grain boundary phase of the ATI 718Plus. According to tests, adding more aluminium results in a rise in yield strength but a steady decline in ductility. When compared to other superalloys like RENE-41 and Waspaloy, the alloy is a better substitute for Inconel 718 because it can tolerate temperatures as high as 704°C [42, 43].

6.2.5 Titanium and Alloys

The titanium alloy is composed of two distinct phases, namely the α-phase and β-phase. The β-phase has a body-centred cubic (BCC) lattice structure and exhibits a moderate tensile yield strength ($\sigma 0.2$ = 860 MPa) while also being highly ductile [54]. On the other hand, the α-phase has a hexagonal close-packed (HCP) lattice structure, which renders it relatively brittle but stronger. Depending on how the phases are arranged, titanium alloys can be divided into five fundamental types: near-α Ti alloys, α+ β Ti alloys, near-β Ti (metastable) alloys, and β-Ti alloys are all types of titanium alloys [54]. The short explanations of different sorts of titanium alloys are as follows:

- **α-Ti alloys**: These alloys normally only exist as the -phase at normal temperatures, with the exception of some varieties of commercially pure titanium (CP-Ti), which may also contain the grain boundary phase. The flexibility of their HCP crystal structure, which offers less slip than BCC structures [54], may be constrained during cold working procedures, particularly at temperatures that are low. In order to prevent the production of coarser grains, hot-working procedures with temperatures below the α-β transition temperature are frequently required for these alloys. The high strain hardening rate of α-Ti alloys should be recognised, as it may limit their formability [55].
- **Near α-Ti alloys**: Adding small amounts of β stabilisers (1% to 2%) to α-Ti alloys can achieve a balance between strength properties, improved creep resistance, and weldability, thereby improving their low-temperature strength and workability. These titanium alloys are classified as α-Ti alloys, which contain both retained and significant amounts of β phases. The addition of β stabilising components can improve their strength, corrosion resistance, and Young's modulus [56]. α-Ti alloys and near α-Ti alloys are highly weldable compared to other titanium grades but cannot be heat treated due to the absence of phase change. However, annealing and stress-relieving techniques can be used on both types of alloys to improve their properties. These alloys are widely used in various applications, such as structural components, airframe skins, gas turbine engine casings, and jet engine compressor blades [57]. At high temperatures (588–811 K), they exhibit strong strength, good notch toughness, acceptable ductility, superior creep, and oxidation resistance [58].
- **α + β Ti alloys**: In α + β Ti alloys, typically 4–6% of β stabilisers are added [54]. These alloys contain both visible α and retained β phases that form a stable microstructure at room temperature, with more retained β phases than in near α-Ti alloys. The most widely used α + β Ti alloy, Ti-6Al-4V (TC4), exhibits three basic α + β microstructures: lamellar, bimodal, and equiaxed [54]. To achieve the desired microstructures and mechanical properties for a specific service environment, α + β Ti alloys can be heat treated at different temperatures. Although most of these alloys can be welded, some may lose their ductility in the weld zone. α + β Ti alloys, particularly those with a high aluminium equivalent, possess high strength and good toughness. However, the existence of the -phase, which has a greater diffusivity and more slip systems than most α-Ti alloys, might influence these alloys' weldability and creep strength, which can make them difficult to cold-form [59].
- **Near β-Ti (metastable β) alloys**: When the martensitic transition point is below room temperature but not by much enough to induce it to do so, alloys with 10%–15% stabilisers show a metastable phase [54]. These metastable phases can undergo an ageing treatment to create a fine Widmanstätten structure in the matrix [54]. These alloys have great hardenability, excellent

forgeability, and high strength and toughness over a broad temperature range. However, they often exhibit lower heat resistance and are susceptible to microstructure instability.

- **β-Ti alloys**: β-Ti alloys with higher stabilising element contents, often up to 30%, have a stable BCC lattice structure at ambient temperature [54]. These alloys have a low concentration of -stabilising components to improve mechanical strength, corrosion resistance, and thermal stability, including creep resistance at intermediate temperatures. In general, β-Ti alloys have stronger and more fatigue-resistant properties than α-Ti alloys do. After being treated with a solution, they also have a wider range of malleable temperatures and show strong cold-forming characteristics. However, sufficient eutectoid elements in -Ti alloys can cause hardenability and embrittlement due to the precipitation of brittle compounds under prolonged heating conditions [54]. Because of this, β-Ti alloys are not widely used and only make up a small percentage of the titanium used in the aerospace sector. At high temperatures, the -phase is also prone to substantial self-diffusive activity and creep. Due to their high density, refractory properties, and poor ductility at room temperature, β-Ti alloys are typically employed in applications that call for great refractoriness and corrosion resistance [54].

β-Ti and near β-Ti alloys have excellent mechanical properties from ambient to intermediate temperature settings and are compatible with heat treatment and welding. These alloys have improved fracture toughness at the same strength level as α + β Ti alloys. The relatively soft phase, which has sufficient slip networks for dislocation motion following solid solution treatment, is responsible for their remarkable cold formability. Alloying elements affect the α-β transus temperature, which is the temperature at which a phase completely changes into another phase or vice versa [60]. The α stabilisers, such as Al, C, Ga, Ge, La, and Ce, increase the α-β transus temperature, whereas β stabilisers, such as Mo, Cr, Nb, Co, Ni, and Mn, decrease it. Sn and Zr have little impact on the phase stability of titanium alloys, while O, N, V, and Fe can stabilise both phases [61]. Due to its excellent solid solubility in both the and phases, which makes it a very efficient strengthener for the solid solution process, Al is the most desired alloying element in titanium alloys. The density of titanium alloys decreases as the Al content increases while the β transus increases. The properties and characteristics of titanium alloys can be significantly influenced by various stabilising components, resulting in a wide range of applications [62]. The β stabilising components and the final microstructure of titanium alloys are what give each type of Ti alloy its unique microstructure and mechanical capabilities. The literature [63–66] provides information on how alloying elements affect the mechanical properties of titanium alloys. The α-β transus temperature ultimately determines the final microstructure and mechanical properties of deposited components, and it also affects the microstructure evolution during the WAAM process, making it essential for the phase transition of titanium alloys. Table 6.5 lists several titanium alloys used in WAAM, and Figure 6.14 shows the variation in the microstructure of these alloys deposited by WAAM.

TABLE 6.5
Different Ti-Alloy Filler Wires for WAAM Process (Composition % w/w)

AWS Nomenclature	C	O	N	H	Fe	Al	V	Pd	Ni	Mo	Ti	Others	Reference
ERTi-2	0.03	0.08–0.16	0.015	0.008	0.12	—	—	—	—	—	99.6	—	AWS A5.16
ERTi-11	0.03	0.03–0.10	0.012	0.005	0.08	—	—	0.12–0.25	—	—	Bal.	—	AWS A5.16
ERTi-12	0.03	0.08–0.16	0.015	0.008	0.15	—	—	—	0.6–0.9	0.2–0.4	Bal.	—	AWS A5.16
ERTi-33	0.03	0.08–0.16	0.015	0.008	0.12	—	—	0.01–0.02	0.35–0.55	—	Bal.	0.1–0.2 (Cr)	AWS A5.16
ERTi-30	0.03	0.08–0.16	0.015	0.008	0.12	—	—	0.04–0.08	—	—	Bal.	0.2–0.8 (Co)	AWS A5.16
ERTi-32	0.03	0.05–0.10	0.012	0.008	0.20	4.5–5.5	0.6–1.4	—	—	0.6–1.2	Bal.	0.6–1.4 (Sn) 0.6–1.4 (Zr) 0.06–0.14 (Si)	AWS A5.16
ERTi-5	0.05	0.12–0.20	0.03	0.015	0.22	5.5–6.75	3.5–4.5	—	—	—	Bal.	—	AWS A5.16
ERTi-23	0.03	0.03–0.11	0.012	0.005	0.20	5.5–6.5	3.5–4.5	—	—	—	Bal.	—	AWS A5.16
ERTi-21	0.03	0.1–0.15	0.012	0.005	0.2–0.4	2.5–3.5	—	—	—	14.0–16.0	Bal.	2.2–3.2 (Nb) 0.15–0.25 (Si)	AWS A5.16
ERTi-36	0.03	0.06–0.12	0.02	0.003	0.03	—	—	—	—	—	Bal.	42.0–47.0 (Nb)	AWS A5.16
ERTi-19	0.03	0.06–0.1	0.015	0.015	0.20	3.0–4.0	7.5–8.5	—	—	3.5–4.5	Bal.	5.5–6.5 (Cr) 3.5–4.5 (Zr)	AWS A5.16

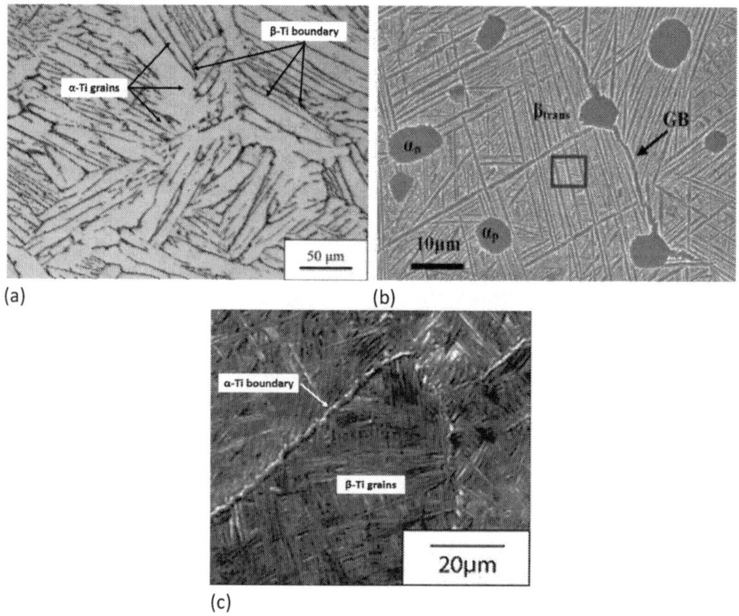

FIGURE 6.14 Microstructure of (a) α-Ti alloy, (b) α+β-Ti alloy, and (c) β-Ti alloy ([a] adapted from Li et al. [67], under CC-By-ND 4.0; [b] adapted from Tan et al. [68], under CC-By 4.0 license; and [c] adapted from Kolli et al. [69], under CC-By 4.0 license)

6.2.6 Refractory Metals

Refractory metals are a class of metals with exceptionally excellent mechanical qualities, corrosion resistance, and wear resistance. Their usual melting point is above 1,800°C. Because of their distinct thermal and structural characteristics, four refractory metals – tungsten (W), tantalum (Ta), niobium (Nb), and molybdenum (Mo) – are frequently utilised in the WAAM process.

6.2.6.1 Tungsten (W)

The tungsten has the highest melting point of any metal, which is around 3,422°C [70]. It is extremely resistant to oxidation and wear and has a high density and mechanical strength. In the WAAM process, tungsten is frequently utilised to create high-strength components, including those used in the aerospace and defence sectors, where toughness and wear resistance are essential. The fabrication of tungsten tiles for the nuclear fusion diverter can be considerably improved by utilising AM, allowing for fine-tuning of their properties based on the desired use [71]. Moreover, WAAM can make it easier to create functionally graded structures with custom compositions and attributes, which can greatly enhance component performance and structural integrity [72]. The maintenance and repair of tungsten plasma-facing components can be done in situ thanks to the extraordinary flexibility and modularity of WAAM techniques, such as robot-based manipulation, which significantly lowers operating costs for nuclear fusion reactors. Because of their poor weldability, however, relatively little study has been done on tungsten alloys using the WAAM technique. Its low ductile to brittle transition temperature (200°C) [70], high melting point, and enormous affinity towards ambient gases up to a wide range of temperatures are the causes of its inferior weldability. Because of these restrictions, the additively produced tungsten alloys are inevitably porous and cracked. Despite the fact that several research studies have used laser and powder in AM of tungsten, the literature has mostly discussed problems with micro-cracks and porosity, leading to incomplete deposits. As a result, the structural integrity needed for nuclear applications is hampered, which could result in failure and a shorter lifetime when exposed to neutron radiation. Even with the substrate being preheated to 400°C, substantial cracking and porosity have been seen in laser powder bed fusion (LPBF) [70]. The balling phenomena have been brought on by the high surface tension and high viscosity of tungsten, which cause slow wetting and spreading speeds. Notwithstanding the difficulties with the material being deposited, Marinelli et al. [70, 71] were able to successfully deposit a rather large-scale component utilising WAAM. In this instance, however, the deposition procedure produced structures devoid of internal fissures, and the layer dimensions remained constant throughout. This enabled the successful manufacture of a somewhat large-scale component measuring 210 × 75 × 12 mm^3 [71]. The fact that such a component was successfully deposited shows the WAAM technique's potential for overcoming the difficulties involved in tungsten deposition. Moreover, it implies that the method can be improved to generate high-quality tungsten components of varied sizes, possibly for use in a variety of applications [71]. There are various types of tungsten wire for WAAM, including magnetic grade and non-magnetic grade. Table 6.6 lists the types of tungsten wire that are readily accessible for WAAM applications. The non-magnetic alloy contains binders like Ni or Cu that significantly increase the basic tungsten alloy's ductility and toughness. Likewise, Fe is used as a binder in magnetic-type alloys for a related function.

6.2.6.2 Tantalum (Ta)

Tantalum has a high melting point of roughly 2996°C [73] and is a rare, dense, and corrosion-resistant metal. It is frequently utilised in the WAAM process to create components for nuclear reactors and medical implants that need to be extremely

TABLE 6.6
Type of Tungsten Alloys for WAAM Process

Name	W (w/w %)	Type	Reference
Unalloyed W	99.99	Non-magnetic	[70–72]
HA190	90	Non-magnetic, class 1	ASTM B777-15
HA1925	92.5	Non-magnetic, class 2	ASTM B777-15
HA195	95	Non-magnetic, class 3	ASTM B777-15
HE390	90	Magnetic, class 1	ASTM B777-15
HE3925	92.5	Magnetic, class 2	ASTM B777-15
HE395	95	Magnetic, class 3	ASTM B777-15
HE397	97	Magnetic, class 4	ASTM B777-15

strong and durable. The metallic element tantalum has special physical and chemical characteristics that make it the perfect material for high-temperature applications [74]. Because of its high melting point, it can tolerate high temperatures without melting or deforming. Furthermore, the surface of tantalum spontaneously develops a highly persistent oxide layer that renders it remarkably resistant to corrosive and chemical attacks at high temperatures [75]. Because of this, it is a very reliable material for usage in challenging conditions where other materials can fail or fast decay. The ductility of tantalum is another significant quality. Tantalum is a very ductile metal, in contrast to other refractory metals like tungsten and molybdenum, which are brittle and prone to fracture at room temperature [75, 76]. This makes it a highly adaptable material for a variety of applications because it can be easily moulded and sculpted into complex forms. Tantalum is used in a variety of firms and industries, including aerospace, defence, electronics, and chemical manufacture, as a result of its special mix of qualities. It is employed to create pieces like heat exchangers, boiler parts, and vessels for chemical reactions, among other things [74, 75]. It is also a highly sought-after material for use in the creation of medical implants, such as artificial joints and dental implants, where biocompatibility and long-term dependability are crucial considerations. This is due to its outstanding chemical stability and resilience to high temperatures. A fine equiaxed microstructure is preferred in the majority of tantalum-using industries to enhance mechanical characteristics and achieve complete isotropy [71, 74]. Yet, when processed by AM, tantalum and other metallic materials frequently have distinctive microstructural characteristics, such as massive columnar grains that cause anisotropy [74]. While there are many approaches to achieving microstructural improvement in AM structures, some techniques, such as changing process parameters or alloy chemistry, may not always be acceptable for preserving process stability and alloy purity. However, it has been discovered that eliminating columnar grains and strong fibre patterns in materials like Ti-6Al-4V can be accomplished by causing plastic deformation inside the deposited layer and then the following deposition of a new layer [77]. Tantalum microstructural refinement has already been the subject of research into the effects of rolling, high-pressure torsion, and equal-channel angular pressing, with the production of ultrafine and nanostructured tantalum components using severe plastic deformation steps on a typical coarse-grained structure [78–80]. Since a strong roughness would result in an irregular erosion rate, a tantalum target that has a fine microstructure is preferred. Depending on the number of impurities and strain/deformation, the recrystallisation temperature of unalloyed polycrystalline tantalum typically ranges from 900°C to 1,450°C [75, 79]. The recrystallisation temperature typically rises when there are more impurities present. Recrystallisation normally requires strain levels between 1.2 and 4.6, and the temperature required for each strain level varies. For instance, samples with a strain of 1.2 recrystallise at 1,100°C, while those with a strain of 3.48 do so at 900°C [77, 79, 80]. Components made of arc-melted tantalum can be substantially deformed by metalloids like carbon, nitrogen, and oxygen. Tantalum is unique among BCC metals in that it has significant interstitial solubility, which means that it doesn't tend to segregate at the grain boundaries. Its improved ductility at room temperature is a result of this distinctive feature [73]. Nevertheless, the purity, temperature, and strain rate of tantalum are significantly correlated with its mechanical properties [78–80], with interstitial elements like nitrogen and oxygen having a significant influence on its hardness and yield strength. Also, the manufacturing process used greatly affects the mechanical characteristics of the item. For instance, commercially available undiluted tantalum components made by powder metallurgy and electron beam melting share similar mechanical characteristics that are primarily influenced by grain size [76, 77, 79]. Tantalum wire of technical grade usually adheres to the ASTM B365-12 standard, whereas medical grade wire follows ASTM F560 or ISO 13782. The wire is commonly provided in either annealed or un-annealed form, either on spools or as straightened tantalum wires. The grades R05200 and R05400 come under the medical grade wires, whereas Ta-Nb and Ta-W alloy come under the industrial grade wires, which may be used for the WAAM process.

6.2.6.3 Niobium (Nb)

Because of its high melting point (2,468°C), extraordinary corrosion resistance, and outstanding biocompatibility, niobium (Nb) is a material that is receiving more and more attention in the high-temperature and biomedicine fields [81, 82]. The surface of Nb is typically oxidised to Nb_2O_5 when exposed to oxygen, which makes it a superb corrosion-resistant material even at high temperatures and stable against corrosive solutions like sodium hydroxide, weak sulphuric acid, hydrochloric acid, or nitric acid [83, 84]. In contrast to more widely used medical metals like tantalum, titanium, and stainless steel, research on Nb began much later [85, 86]. However, much clinical research carried out over the previous three decades has shown that Nb possesses great osteogenic qualities by boosting the activity of alkaline phosphatase to induce calcification in addition to having high biocompatibility and minimal cytotoxicity [87]. Nb is a BCC refractory metal, but due to its sensitivity to oxygen, even a small amount can significantly alter its mechanical characteristics [85–87]. Pure Nb's use is also severely constrained by the high cost of raw sources, its high melting point, and its affinity for oxygen. For orthopaedic implants, which require high structural freedom and customisation for each individual, conventional methods face great challenges. Despite these limitations, Nb and its alloys are gaining attention in the AM field, particularly in WAAM processes. Nb and its alloys have been successfully fabricated using this process, with the added benefits of high efficiency, excellent control of material properties, and reduced waste generation [88]. Furthermore, the ability to customise the implants according to the individual's needs and the use of Nb's exceptional biocompatibility and osteogenic properties make WAAM a promising technology for the fabrication of orthopaedic implants. Nb doesn't readily absorb neutrons in a nuclear reactor because it has a low thermal neutron capture cross-section [81, 83]. Because of this, Nb can be used in some nuclear applications, such as the cladding of nuclear fuel rods. Superconducting magnets for fusion reactors and particle accelerators are among the various uses of Nb in the nuclear sector. On the other hand, Nb is a desirable material for the jewellery business since it can be anodised and coloured [81]. Nb's anodised layer is very stable, and the resulting colours are vibrant and long-lasting. This feature, along with Nb's hypoallergenic properties, makes it an excellent material for creating colourful and safe jewellery. Nb wires are available in two grades (ASTM B392), reactor grade and commercial grade, both of which are primarily available in their pure form. The reactor-grade unalloyed Nb is categorised as R04200-type 1, while the commercial-grade unalloyed Nb is classified as R04210-type 2. The R04210 alloy is commonly used for medical applications. For commercial use, the R04261-type 4 alloy, which contains 1% Zn, is also available.

6.2.6.4 Molybdenum (Mo)

Molybdenum (Mo) has a remarkably high melting point (2623°C) amongst the group of refractory metals and other extraordinary qualities that make it essential in a wide range of applications, including but not limited to the aerospace [89], nuclear [90], and electronics sectors [91]. Mo and its alloys are challenging to treat due to their brittleness and hardness. Currently, Mo structures are often made via powder metallurgy. Crushed, sintered, and thermo-mechanically treated powders are used in this process to create mill products like rods and plates. Then, these products can be manufactured, assembled, and machined to create specialist components. There aren't many effective techniques to remove the interior pores between the powders, which limits the powder metallurgy technique's ability to produce large-scale and tailored goods [91]. The introduction of AM has created new opportunities for making materials that are difficult to process, such as complicated structures, large-scale parts, and alloys of aluminium that don't shatter [92, 93]. This offers the possibility of mass-producing complex Mo structures without the use of moulds. Due to fabrication-related flaws such as cracks and pores, the AM of Mo continues to be very difficult [91]. The performance of the structure can be severely hampered by these flaws, necessitating thermodynamic processing [94]. Researchers have largely focused on fixing these issues with Mo synthesised by selective laser melting (SLM), selective electron beam melting (SEBM), DED, and other techniques in order to manufacture high-quality Mo structures [94–100]. The main obstacles to fabricating Mo parts using AM are the absence of fusion caused by Mo's high melting point, the pores produced by impurity elements [97], the high stress associated with AM [101], and the cracks caused by Mo's innate brittleness [91, 95]. Thus, researchers have investigated a variety of strategies, including optimising the manufacturing process, lowering the impurity concentration, adding alloying elements, and other techniques [98–100] in order to obtain high-density and crack-free Mo parts. Low power produced porous samples with a maximum relative density of only 82.5%, according to Faidel et al. [95]. They asserted that lowering scanning speed alone was insufficient for higher-density parts and that stronger laser power and lower layer height were required. Similar findings were made by Braun et al. [97], who discovered that oxygen segregation at grain boundaries during SLM resulted in Mo oxide production, which triggered crack nucleation and propagation as well as the creation of pores. They stressed the possibility of preventing these events by removing contaminants, particularly oxygen. John et al. [98] discovered that using a Mo substrate requires a high laser power of 6,000 W using DED. When reducing laser power, preheating the substrate led to substantial residual stress and fissures between the sample and substrate. While flaws were decreased by using a stainless-steel substrate, cracks and pores continued to occur along grain boundaries. In order to generate high-density Mo structures and overcome the cracking issue, Kaserer et al. [94] alloyed unalloyed

Mo with a small quantity of carbon during SLM. By removing oxygen, this method produced a relative density of 99.7 0.2%, which is higher than the predicted density of Mo-0.45wt%C (10.11 g/cm^3). Wang et al. [91, 100] enhanced packing density and fluidity by improving powder quality to produce unalloyed Mo with relative densities of 99.1% and 97.7%, respectively, using SLM and WAAM. According to Higashi et al. [99], SLM that had a volumetric energy density greater than 150 J/mm^3 produced a relative density of more than 99%. They linked pores to the behaviour of Mo during oxidation. Mo components with a relative density of 99.5% were created by Rebesan et al. [91] via laser powder bed fusion; however, their mechanical characteristics were anisotropic and inferior to those of annealed Mo samples. Despite ongoing efforts to reduce pore and crack formation, little research has been done on wire arc AM and the mechanical characteristics of Mo produced using AM methods. Few studies that address the flaws, metallurgical, and mechanical qualities have been reported in this respect. With the help of WAAM, Wang et al. [91, 100] were able to create parts for Mo that were devoid of cracks and had a high density and typical size. When the microstructure and mechanical characteristics of samples that had been heat treated and those that had not were compared, it became clear that the heat treatment had impaired the mechanical properties by causing the grain and sub-grain boundaries to stop hardening. The strength and ductility of the as-deposited and heat-treated specimens were comparable at high temperatures. Yet, during high-temperature tests, MoO_3 shards could be seen on the fracture surface. The internal and exterior flaws in unalloyed Mo alloys deposited using the WAAM technique were analysed by Qi et al. [102]. Findings showed that the exterior outline changed wildly, and the sample diameter rose logarithmically as frequency increased. Moreover, internal flaws grew more vertical, indicating a change in the morphology of the molten pool's solid-liquid interface. Due to the extremely high temperature that was maintained at the centre of the molten pool under the condition of 20 Hz, which caused the molten metal to evaporate constantly, cavity growth happened in the centre from bottom to top. The development of WAAM and a deeper comprehension of the dynamics of the molten pool during production can both benefit from the study of frequency characteristics.

Unalloyed Mo wires have traditionally been employed in WAAM applications. However, alternative alloys are also available for this purpose, according to the ASTM B387 standard. Mo wires are known as 360, 361, and 365 when they are unalloyed and 363 and 364 when they have a small amount of titanium added. The 366 alloy, which has a considerable percentage of tungsten (27%–33%), is an additional choice.

6.2.7 Other Metals

6.2.7.1 Nickel Aluminium Bronze (NAB)

The average composition of NAB is 8%–10% aluminium, 3%–5% nickel, and the rest copper [103, 104]. The strength and corrosion resistance of the copper alloy is increased by adding nickel and aluminium. Because of its exceptional qualities and capacity to withstand corrosion in saltwater conditions, NAB is frequently used in marine and aeronautical applications [105]. The alloy also has a reputation for having a high level of wear resistance, which makes it a good choice for parts that are prone to significant wear and tear. A number of researchers have employed WAAM to create NAB components for naval applications [106, 107]. Although stress relaxation occurs at the same time as the expansion of the phases, they have observed that NAB alloys are sensitive to heat treatments [103, 107]. Although these heat treatments need to be improved for laser-based AM, they have been shown to be effective in creating good mechanical characteristics for NAB material deposited using wire arc DED [107] when heated to 675°C for six hours. When compared to traditional cast or wrought materials, wire arc DED depositions produce a consistent Widmanstätten-microstructure with little anisotropy, according to numerous investigations. The high reflectivity of NAB has also been effectively overcome via laser wire/powder DED and L-PBF techniques, resulting in high-density constructions despite NAB's high reflectivity. However, the high cooling rates involved in laser processing lead to the formation of a microstructure made of martensitic β′ and/or Widmanstätten-α, which improves strength at the expense of ductility [106, 107]. Furthermore, when compared to traditional cast and wrought materials, laser-based processing methods have shown that NAB has improved corrosion resistance. Even though there hasn't been much research done to analyse electron beam-powder bed fusion (EB-PBF) NAB, the preliminary findings show that it has better strength and ductility than wrought material [104]. Because of casting porosity, conventional casting methods for NAB components frequently produce coarse microstructures and reduced mechanical characteristics. The characteristics of cast NAB components have been improved using a variety of processes, although their applications are restricted [107]. These procedures include heat treatment, friction stir processing, and fusion welding. Nevertheless, layer-by-layer structure creation using WAAM has demonstrated its ability to create components with close-to-net shapes, fine microstructures, and mechanical properties that are comparable to those of as-cast NAB alloy [103–107].

6.2.7.2 Ni-Ti

Several researchers have observed using WAAM to create different intermetallic compounds (IMCs), such as Fe-Al [108], Ti-Al [109], etc. Nevertheless, due to its severe brittleness, wired IMC cannot be produced, so individual wires from each element are used via a twin wire system to carry out in situ fabrication via WAAM [108, 109]. Due to its inherited ductility and shape memory effect, Ni-Ti is one of the most significant IMCs, both architecturally and functionally. An example of a shape memory alloy comprised of nickel and titanium is Ni-Ti, sometimes referred to as nitinol. It is a unique material that has the ability to recover its original shape after deformation, a property known as the shape memory effect (SME), and also can exhibit superelasticity, where it can undergo large reversible deformation without permanent damage. The unique properties of Ni-Ti are due to a reversible phase transformation that can occur in the material between two crystal structures, known as austenite and martensite. When Ni-Ti is heated above a certain temperature, called the austenite finish temperature (A_f), it transforms from martensite to austenite [110]. When it is cooled below the austenite start temperature (A_s), it transforms back to martensite. This transformation can occur reversibly, allowing for shape memory and superelasticity [111]. Iron-aluminium, copper-aluminium, and titanium-based alloys have all been produced using WAAM on a large scale; however, nickel-titanium (Ni-Ti) shape memory alloys have not been substantially deposited using WAAM until recently. Ni-Ti alloys are distinctive materials with shape memory and superelastic characteristics, which make them perfect for a variety of applications, such as robotics, aircraft, and medical implants [112]. Yet, it can be difficult to manufacture Ni-Ti alloys with intricate geometries and structures using conventional manufacturing techniques. The WAAM procedure was used by Wang et al. [110] to create a Ni-rich Ni-Ti alloy. The Ni-Ti alloy with a 53.5 at % Ni content was deposited on the Ti substrate after the author melted pure Ti and Ni wires. The sample structure that was produced, though, was quite heterogeneous and included precipitates that were rich in nickel, including Ni_4Ti_3 and Ni_3Ti. The differential scanning calorimetry study of the martensitic transformations in various regions of the Ni-rich Ni-Ti sample revealed the presence of very weak heat release and heat absorption peaks, indicating that the martensitic transformations were partially suppressed, leading to poor functional behaviour. The use of elemental Ti and Ni wires rather than Ni-Ti alloy wire may have contributed to the heterogeneous nature of the WAAM sample prepared for the investigation [111]. To solve this issue, Zheng et al. [113] employed a GTAW-based WAAM technique with Ni 49.5 at. % Ti wire as the feedstock to create a 5-layered Ni-Ti sample on a Ni-Ti substrate. The structure, the martensitic transformation, and the mechanical behaviour in various layers were not found to be correlated, despite the fact that superelasticity and the martensitic transformation were explored in this sample. Yet, in order to comprehend how the layer structure affects the functional behaviour of Ni-Ti alloy components made by WAAM, such relations are essential. In a different investigation, Wang et al. discovered that the Ni-Ti alloy created by WAAM had a significant amount of Ni_4Ti_3 precipitates that were nanoscale in size. The precipitates were disposed inside the alloy at an angle of 60°, taking the form of lenses. Also, it was investigated how the size of the Ni_4Ti_3 precipitates related to the deposition current. The findings demonstrated that the average width of precipitates rose from 96.7 nm to 164.1 nm with an increase in deposition current from 80 A to 120 A. WAAM was used to create Ti_2Ni precipitates in the Ni-Ti alloy. The macroscopic characteristics of the Ni-Ti alloy were significantly influenced by the various microstructures. The stability of the Ni-Ti alloy produced by WAAM in the superelastic cycle was assessed by Zeng et al. [114]. Under the pre-strain of 6%, 1.13% residual strain was produced in the first cycle, and as the number of cycles grew, the unrecoverable strain eventually stabilised at 2.73%. This was a result of the Ni-Ti alloy's uneven grain sizes and initial residual martensite [110]. All of these investigations came to the conclusion that WAAM can be a practical manufacturing technology for the production of components with near-net shapes for various structural and functional applications. Ni-Ti alloys are used in a variety of industries, including aerospace, automotive, and medical devices. Because of its biocompatibility and distinctive qualities, nitinol is employed in medical applications such as stents, orthodontic wires, and surgical equipment. Because of its small weight and shape memory capabilities, nitinol is employed in actuators, sensors, and other components in the aerospace and automotive industries.

6.3 TYPE OF RAW MATERIAL FOR WAAM

6.3.1 Wire-Type Raw Materials

In WAAM applications, which make use of the MIG/TIG welding principle, wire-type raw materials are frequently used. There are several different diameters of MIG/TIG wires, including 0.6 mm, 0.8 mm, 1 mm, and 1.2 mm. Due to its superior control over layer height and width, the 1.2 mm diameter wire is the one that is utilised in WAAM procedures the most frequently. It's crucial to remember, though, that smaller wire sizes can lead to smoother surfaces and thinner layers. With a 0.6 mm diameter wire, it would take twice as many passes to deposit the same height structure as with a 1.2 mm wire, resulting in additional thermal cycles and a possible loss of mechanical and metallurgical attributes. Additionally, WAAM is known for its high deposition rates, which may be nullified with smaller diameter wires. As a result, researchers and industrialists prefer using 1.2 mm diameter wires to achieve structures with optimum deposition rates and mechanical properties while avoiding excessive thermal residual stress.

Furthermore, the impact of droplets on the molten metal pool should also be considered when selecting the wire diameter for WAAM applications. Thinner wires produce smaller droplets than thicker wires, which can create agitation when they fall into the molten metal pool during deposition. In a study on the effects of droplet impact on the dynamics of molten metal pools, Chen et al. [115] reported simulated findings that show the major influence of droplet impact on welding procedures. Droplet impact changes the surrounding liquid's concave shape and forces it upward, which has an impact on heat transfer, fluid flow, and weld reinforcing. Droplet impact is taken into account, which raises the Peclet number, improves the molten pool's convection and homogeneity, and causes the weld to be taller and thinner. The study concludes that understanding droplet impact is crucial for optimising and precisely controlling welding processes and provides a quantification basis for weld geometry in engineering applications. Therefore, the size of molten droplets should be large enough to create agitation in the melt pool but not that large to create splashes in the molten metal pool, which can lead to a spatter.

6.3.2 Powder-Type Raw Materials

Certain WAAM systems also use powder as a raw material in addition to wire. In most cases, a plasma arc is used as the heat source in powder-based WAAM systems to melt the powder, which is then deposited layer by layer to produce the finished item. The raw material in a powder-based WAAM system is typically metal powder, which is supplied into the system using a hopper or other similar mechanism. An electric current is then applied to a gas (such as argon or helium) flowing through a nozzle to form a plasma arc, a type of high-temperature gas discharge, which is used to heat the powder. The powder is melted using the high-temperature zone produced by the plasma arc. To construct the 3D object, the molten powder is then layer by layer and put onto a support or earlier layer. The procedure is comparable to the wire-based WAAM method, except the system employs powder as its raw material rather than wire.

One of the advantages of the powder-based WAAM process is that it can create objects with greater geometric complexity than the wire-based process, as the powder can be more easily distributed and shaped than a wire. Additionally, the process can be used with a wider variety of materials, as many metals can be obtained in powder form. However, the powder-based WAAM process is less common than the wire-based process, as it requires more sophisticated equipment and can be more challenging to control. Additionally, the plasma arc used in the process can generate hazardous gases and fumes, so appropriate safety measures must be taken. Overall, the choice of raw material for a WAAM system will depend on the specific application and the desired properties of the final object.

6.3.3 Hybrid-Type Raw Materials

The hybrid-type raw materials for the WAAM process are a combination of powder and wired raw materials. For MIG and TIG welding applications, flux-cored wires have been prominently used for two decades. However, it has been incorporated into the WAAM process recently. Flux-cored wires are a type of wire used in the WAAM process. These wires have a hollow core filled with a flux material that is released as the wire melts during the welding process. The flux material creates a shield around the weld area, protecting it from contaminants in the air and reducing the risk of defects. Flux-cored wires are often used in WAAM because they provide several advantages over solid wires. One advantage is that they can be used with a wider range of welding parameters, such as higher current and voltage settings, allowing for faster deposition rates [116]. They can also produce a deeper penetration than solid wires, which is useful for welding thicker materials. There are several types of flux-cored wires available for use in WAAM, with different compositions and characteristics to suit different applications. For example, some wires are designed for use with specific materials such as stainless steel or nickel alloys. Others may have higher deposition rates or provide better weld quality in certain positions, such as overhead or vertical welding. Flux-cored wire arc additive manufacturing (FCWAAM) is a term used by Zhang et al. [117] to describe the introduction of flux-cored wire into the WAAM procedure (FCWAAM). They used duplex stainless steel flux-cored wire for the deposition using the WAAM method, and they were able to produce a homogeneous microstructure with limiting austenite concentration at 37% and 45%, which is considerably less than that of solid wire, which is 74%. Wang et al. [118] created a novel FCWAAM technique to produce parts of TiB and TiC-reinforced Ti6Al4V matrix composites using the FCWAAM. B4C and C powders in the flux core caused in situ reactions that resulted in the formation of a network microstructure and reinforcing. A refined net-basket-dominated (α+β)-Ti matrix and a stable network microstructure formed in the intermediate region according to the results of an investigation into the impact of reinforcing percentage on microstructure and wear resistance. The 10% wt. sample demonstrated poorer wear performance due to decreased ductility and increased cracking, and microhardness increased with reinforcing fraction. In contrast with FCWAAM, the metal-cored WAAM (MCWAAM) process has also been developed recently to incorporate the in situ alloying during the deposition [119–121]. Truetler et al. [119] developed CoCrFeNiV high entropy alloy (HEA) by using metal-cored wire via the WAAM process. They were able to deposit the suggested HEA with a homogeneous microstructure, demonstrating WAAM's potential for creating concentrated complicated alloys. Also, in another work, a seamless and very useful multi-walled component constructed of 1.25Cr-0.5Mo stainless steel was produced using WAAM with metal-cored wire [121]. Tempered martensite was present in the bottom zone of the component's microstructure, while finer microstructures were present in the middle and top zones. The component's Microhardness levels remained consistent throughout. The mechanical characteristics of the multi-walled structure were similar to those of the metal-cored wire, according to tensile and impact tests. The results suggest that MCWAAM is suitable for manufacturing multi-walled structures of 1.25Cr-0.5Mo for industrial applications [121]. However, MCWAAM is quite nascent in the technique, and more research should be conducted on it in the very near future for proper implementation of the technique into the WAAM process. Specifically, the MCWAAM process could help the material scientist in high throughput material development (HTMD) by mixing different material powders altogether and fusing all of them in well agitated molten metal pool created by the electric arc.

Finally, wires with metal or flux cores can both be employed in the WAAM method. Due to its capacity to deliver shielding gas and fluxing agents concurrently, flux-cored wires are frequently employed for WAAM. This helps to prevent oxidation and improves the metallurgical quality of the deposited material. Nonetheless, metal-cored wires have several benefits over flux-cored wires, including decreased emissions of fumes and spatter, increased deposition efficiency, and a need for less post-processing. Metal-cored wires also have better control over the deposition rate, penetration, and heat input, which makes them suitable for producing high-quality and consistent deposits. Overall, the choice between flux-cored and metal-cored wires for WAAM depends on the specific application requirements and the material properties of the workpiece being fabricated.

6.4 TYPE OF SHIELDING GASES FOR WAAM

During the WAAM process, it is crucial to protect the molten pool from atmospheric gases and prevent oxidation. This is where shielding gases come in. Shielding gases are used to provide a protective barrier around the molten pool and ensure the quality of the deposited material. In this section, the importance of shielding gases in WAAM and the different types of gases that are commonly used will be discussed thoroughly. Shielding gases are used in WAAM to prevent atmospheric gases, such as oxygen and nitrogen, from reacting with the molten metal and causing defects in the deposited material. These defects can include porosity, cracks, and lack of fusion, all of which can weaken the final product [122]. Shielding gases also help to stabilise the arc and improve weld quality by reducing spatter and controlling the heat input [123]. Different types of gases are used or may be used for the shielding of arc during the WAAM process, which can be basically grouped into two categories: inert gases and reactive gases. As per ISO 14175:2008 – welding consumables – gases and gas mixtures for fusion welding and allied processes, argon (Ar) and helium (He) are enlisted in inert gases, whereas carbon-di-oxide (CO_2), hydrogen (H_2), oxygen (O_2), and nitrogen (N_2) are included in reactive gases. However, the mixture of any combination can also be used for inert or/and reactive shielding during the WAAM process. Table 6.7 [122, 123] shows the characteristics of different shielding gases for the

TABLE 6.7
Functions and Application of Various Shielding Gases and Blends

Shielding Gas	Chemical Activity	Observation	Suitability: WAAM	Suitability: FCWA-AM	Economy	Grade (1 = Poor, 5 = Excellent)
Ar	Inert	• Good starting arc and arc stability • Widely used in welding • High penetration • Low thermal conductivity • Unstable arc	Spray or pulsed transfer modes are suited for welding practically all metals.	Spray or pulsed transfer modes are suited for welding practically all metals.	Expensive	4
He	Inert	• Used when increased heat input is desired • Improves wetting action, penetration, travel speed • Good thermal conductivity • Good penetration • Unstable arc	Ideal for copper and aluminium to increase heat and reduce porosity.	Not advised since it offers a poor cost-to-application ratio.	Expensive	5
CO_2	Oxidising	• Its addition results in wider and deeper bead penetration • Causes a high amount of spatter • Restricted to short arc metal transfer • Good thermal conductivity • Good shielding effect	Appropriate for low alloy and plain carbon steels, with deoxidised wire, there is a risk of splatter.	Stable arc, smooth spray transfer mode, works as a carburising or decarburising process for carbon steels, stainless steels, and nickel alloys.	Cheap	3

(Continued)

TABLE 6.7 (CONTINUED)

Shielding Gas	Chemical Activity	Observation	Suitability: WAAM	Suitability: FCWA-AM	Economy	Grade (1 = Poor, 5 = Excellent)
N_2	Reducing	• Reacts at high temperature • Its addition stabilises the arc and reduces • Surface tension • Used as a partial component with hydrogen in root shielding of austenitic stainless steels • Low thermal conductivity	It is not advised since it reacts at welding temperatures and can result in weld flaws (crack susceptibility, porosity).	It is not advised since it reacts at welding temperatures and can result in weld flaws (crack susceptibility, porosity).	Cheap	2
H_2	Reducing	• High thermal conductivity • Its addition can increase heat input • Stable and concentrated arc • Good penetration	Cannot be used as pure; the concentration must be less than 10%.	Cannot be used as pure; the concentration must be less than 10%.	Cheap	2
O_2	Oxidising	• Its addition stabilises the arc and reduces surface tension • Low thermal conductivity • Stabilises the arc with Ar	Utilised just as an additive gas to enhance the quality of the microstructure, wetting, arc stability, and weld profile.	Utilised just as an additive gas to enhance the quality of the microstructure, wetting, arc stability, and weld profile.	Cheap	1
$Ar+CO_2$	Oxidising	• Stable arc • Best for carbon steel • Less spatter and fume	Carbon steels, low- and high-alloy steels can be produced with CO_2 levels of 5% to 25%. The arc is stabilised by CO_2 addition. Short-circuiting and spray transfer modes.	Short-circuit and spray transfer modes, less spatter and fume, better out-position, and 10%–25% CO_2 for steels, stainless steels, and nickel alloys.	Moderate	3

$Ar+O_2$	Oxidising	• Stable arc • Suitable for spray mode of metal transfer	Using deoxidised wire and 3%–5% O_2, the addition of O_2 stabilises the arc and is suitable for diverse steels. Mode of transfer is spray.	Not advised since it offers a poor cost-to-application ratio.	Moderate	2
$Ar+N_2$	Reducing	• Highly reactive metals are not recommended • Spatter and porosity • High heat input	Strong but smoother functioning, more controllable arc on Cu than with just N_2, N_2 (up to 2%), and risk of spatter and porosity for duplex stainless steels.	N_2 is not advised as it increases welding voltage, creates porosity, and decreases welding current.	Moderate	1
$Ar+H_2$	Reducing	• Reduced melting efficiency • Chances of hydrogen embrittlement • Spatter and porosity	Although less effective than in GTAW, adding H_2 can speed up melting and boost melting efficiency, higher vulnerability to hydrogen cracking risk.	H_2 should not be used since it damages the majority of process-relevant base metals.	Moderate	1
$Ar+He$	Inert	• Suitable for spray transfer mode • High heat input • Stable arc	Arc movement that is more easily regulated and quieter (20–80 to 50–50). Heat input is increased when He is added to Ar. Mode of transfer is spray.	Not advised since it offers a poor cost-to-application ratio.	Expensive	3
$He+N_2$	Reducing	• Inferior mechanical property • Porosity and spatter • High heat input	In some semi-austenitic stainless steels, adding more than 1% of N_2 to the weld decreases its tensile strength and ductility.	It's not advised since N_2 causes porosity that works against the flow. Raises the welding voltage while lowering the welding current.	Moderate	1

(*Continued*)

TABLE 6.7 (CONTINUED)

Shielding Gas	Chemical Activity	Observation	Suitability: WAAM	Suitability: FCWA-AM	Economy	Grade (1 = Poor, 5 = Excellent)
$Ar+He+CO_2$	Oxidising	• Smooth bead surface • Stable arc • Short-circuit mode of metal transfer	Attractive weld beads. With stainless steel, only a short-circuit transfer mode is available. To prevent the weld from absorbing too much carbon, CO_2 should be kept below 5%.	Not advised since it offers a poor cost-to-application ratio.	Expensive	4
$Ar+He+O_2$	Oxidising	• Stable arc • Less porosity and spatter • Good for reactive metals	The level of porosity in aluminium alloys is decreased by adding He to Ar and O_2 in amounts less than 0.04%.	Not advised since it offers a poor cost-to-application ratio.	Expensive	4
$Ar+Co_2+N_2$	Redox	• Stable arc • Not recommended for highly reactive metals • Less porosity and spatter	Stainless steels that can be used for transfer welding, short-circuiting, and globular, spray, or pulsed spray (e.g., >90% Ar, 10% CO_2, 10% N_2).	High risk of flux agent reaction, not advised.	Cheap	3
$Ar+Co_2+O_2$	Oxidising	• Not recommended for highly reactive metals • Suitable for steels • Stable arc	5% O_2 + 15% CO_2 for welding deoxidised wire on steels. Short-circuit, spray, and pulsed transfer modes. Arc stability is slightly improved compared to Ar/CO_2.	Not advised since it offers a poor cost-to-application ratio.	Expensive	3

$Ar+Co_2+H_2$	Redox	• Stable arc • Smoother bead • Less spatter and porosity	The WAAM of Inconel 625 alloys is more stable in stainless steels when CO_2 is added in small amounts to combinations of Ar + H_2 or Ar + He. Transfer modes: spray, pulsed, and short circuit.	High risk of flux agent reaction, not advised.	Moderate	3
$Ar+Co_2+He+H_2$	Redox	• Stable arc • High deposition rate • Smoother bead • Less spatter and porosity	Alloys and nickel. Reduced oxides. Enhanced weld profile and deposition rates. Fantastic for pulsing. More stable arc than Ar or Ar + CO_2.	High risk of flux agent reaction, not advised.	Expensive	3
$Ar+Co_2+He+O_2$	Oxidising	• High deposition rate • Smoother bead profile • Less spatter	High current. Mild steel, high strength low alloy (HSLA) steel. High productivity. Similar reliance on cored wire for electrode extension.	High current. Mild steel, HSLA. High productivity. Similar reliance on cored wire for electrode extension.	Expensive	3

deposition purpose. Amongst the available sources of shielding gases, the Ar stands out to be a perfect candidate in terms of economy, availability, and influence on the properties of deposited material. When it comes to choosing the appropriate shielding gases for WAAM, there are several factors to consider. These factors include the type of material being deposited, the type of wire electrode being used, the desired weld properties, and the cost and availability of the shielding gas.

Here are some general guidelines for selecting shielding gases for WAAM:

- For ferrous materials, such as steel or stainless steel, a mixture of argon and CO_2 is commonly used as a shielding gas. The ratio of argon to CO_2 can vary depending on the specific material being welded and the desired properties.
- For non-ferrous materials, such as aluminium or titanium, pure argon or a mixture of argon and helium is commonly used as a shielding gas.
- For high-strength materials, such as nickel alloys, a mixture of argon and helium is often used to provide the necessary heat input while minimising distortion.
- For some specific wire electrodes, such as metal-cored wires or flux-cored wires, specific shielding gas mixtures may be recommended by the manufacturer.
- The flow rate of the shielding gas should also be considered. Generally, a flow rate of 12–20 litre/min is appropriate for most WAAM applications.

It is important to note that the choice of shielding gas can have a significant impact on the quality of the weld, including its appearance, strength, and porosity. Therefore, it is recommended to consult with experts and perform appropriate testing to determine the optimal shielding gas for a specific application.

6.5 SUMMARY

WAAM, a relatively new method, has gained popularity in recent years because of its versatility and affordability. Through this process, metal layers are carefully deposited to produce 3D objects. WAAM offers a variety of advantages over traditional manufacturing processes, including the ability to produce complex geometries, reduced material waste, and shortened lead times.

The selection of material is one of the main aspects that affect a WAAM process's success. Steel, aluminium, titanium, nickel alloys, copper alloys, and other metals and alloys can all be deposited using WAAM. These materials can all be used for various applications because of their distinctive individual characteristics. One of the materials most frequently utilised in WAAM is steel. It provides excellent strength, longevity, and processing simplicity. Depending on the requirements of the particular application, many types of steel, including mild steel, high-strength steel, and stainless steel, can be employed. Another common material for WAAM is aluminium. It has good thermal conductivity, a high strength-to-weight ratio, and great corrosion resistance. Yet, due to its low melting point and strong heat conductivity, aluminium can be difficult to produce. High-strength titanium is frequently utilised in aerospace and medical applications. It has good biocompatibility, low density, and corrosion resistance. Due to its high reactivity with oxygen and nitrogen, processing it can be difficult. In high-temperature applications like gas turbines and jet engines, nickel alloys are employed. They provide outstanding high-temperature strength and corrosion resistance. Nonetheless, because of their high melting point and heat conductivity, they can be difficult to process. WAAM also makes use of copper alloys like bronze and brass. They have good corrosion resistance, high electrical and thermal conductivity, and appealing looks. Yet, because of their high heat conductivity and propensity to oxidise, they can be difficult to process.

Wire is often used as the primary raw material in the WAAM process. Steel, titanium, aluminium, nickel alloys, and other materials used in the industrial sector are just a few of the materials that can be used to make this wire. The final 3D item is produced by melting wire that is fed via a wire feeder and depositing it layer by layer. Certain WAAM systems also use powder as a raw material in addition to wire. In most cases, a plasma arc is used as the heat source in powder-based WAAM systems to melt the powder, which is then deposited layer by layer to produce the finished item. The wire-based process is more common than the powder WAAM procedure. Additionally, two types of wires have begun to be employed in the WAAM process: flux-cored and metal-cored wires. Flux-cored wires have hollow cores that are filled with a flux substance during the welding process to form a shield around the weld region, shielding it from impurities and lowering the possibility of flaws. For faster deposition rates and higher penetration, metal-cored wires, on the other hand, feature a solid core loaded with metal powder that is released as the wire melts. With WAAM, the particular application and required attributes will determine whether to use flux-cored or metal-cored wires.

Shielding gas is a crucial part of the WAAM process because it guards the metal molten pool from oxidation and contamination while it is being deposited. The material being used and the required qualities of the finished product influence the choice of shielding gas. Because of its inertness and efficiency against oxidation and contamination, argon is the most often utilised shielding gas in WAAM. As a co-shielding gas with argon, helium is occasionally utilised to improve heat transfer and weld penetration. Aluminium and magnesium alloys are treated with nitrogen to prevent oxidation and enhance surface quality. Carbon dioxide can be used as a shielding gas for steel to increase the deposition rate, while oxygen is rarely used due to its tendency to cause oxidation and contamination of the molten pool.

ACKNOWLEDGEMENT

The authors would like to acknowledge the Department of Science and Technology-Science and Engineering Research Board (DST-SERB) for the financial support (sanction no. ECR/2018/001250) to carry out the research activities.

CONFLICT OF INTEREST

There is no potential conflict of interest.

REFERENCES

1. Wu B, Pan Z, Ding D, et al. (2018) A review of the wire arc additive manufacturing of metals: Properties, defects and quality improvement. *J Manuf Process* 35: 127–139.
2. Colegrove P (2010) *High Deposition Rate High Quality Metal Additive Manufacture Using Wire + Arc Technology*. Cranfield University
3. Bekker ACM, Verlinden JC, Galimberti G (2016) Challenges in assessing the sustainability of wire + arc additive manufacturing for large structures. *Solid Freeform Fabrication 2016: Proceedings of the 27th Annual International Solid Freeform Fabrication Symposium – An Additive Manufacturing Conference, SFF 2016 1*, 406–416.
4. Li JLZ, Alkahari MR, Rosli NAB, et al. (2019) Review of wire arc additive manufacturing for 3d metal printing. *Int J Autom Technol* 13. https://doi.org/10.20965/ijat.2019.p0346
5. Lambiase F, Scipioni SI, Paoletti A (2022) Accurate prediction of the bead geometry in wire arc additive manufacturing process. *Int J Adv Manuf Technol* 119. https://doi.org/10.1007/s00170-021-08588-w
6. Eyercioglu O, Atalay Y, Aladag M (2020) Evaluation of overhang angle in tig welding-based wire arc additive manufacturing process. *Int J Res – Granthaalayah* 7. https://doi.org/10.29121/granthaalayah.v7.i10.2019.393
7. Wang Z, Zhang Y (2021) A review of aluminum alloy fabricated by different processes of wire arc additive manufacturing. *Medziagotyra* 27. https://doi.org/10.5755/j02.ms.22772
8. Vishnukumar M, Pramod R, Rajesh Kannan A (2021) Wire arc additive manufacturing for repairing aluminium structures in marine applications. *Mater Lett* 299. https://doi.org/10.1016/j.matlet.2021.130112
9. Yang Q, Xia C, Wang H, et al. (2022) Microstructure and MECHANICAL PROperties of TiB2/AlSi7Mg0.6 composites fabricated by wire and arc additive manufacturing based on cold metal transfer (WAAM-CMT). *Materials* 15. https://doi.org/10.3390/ma15072440
10. Wang C, Suder W, Ding J, Williams S (2021) The effect of wire size on high deposition rate wire and plasma arc additive manufacture of Ti-6Al-4V. *J Mater Process Technol* 288. https://doi.org/10.1016/j.jmatprotec.2020.116842
11. Wang C, Suder W, Ding J, Williams S (2021) Wire based plasma arc and laser hybrid additive manufacture of Ti-6Al-4V. *J Mater Process Technol* 293. https://doi.org/10.1016/j.jmatprotec.2021.117080
12. Wang K, Sun Z, Liu Y, Lv Y (2020) Investigation of microstructure and mechanical performance of in738lc superalloy thin wall produced by pulsed plasma arc additive manufacturing. *Materials* 13. https://doi.org/10.3390/MA13183924
13. Paul AR, Mukherjee M, Raja M, et al. (2022) Development of near homogeneous properties in wire arc additive manufacturing process for near-net shaped structural component of low-carbon steel. *Proc Inst Mech Eng C J Mech Eng Sci* 236: 3497–3511. https://doi.org/10.1177/09544062211045489
14. Xin H, Correia JAFO, Veljkovic M, et al. (2021) Probabilistic strain-fatigue life performance based on stochastic analysis of structural and WAAM-stainless steels. *Eng Fail Anal* 127. https://doi.org/10.1016/j.engfailanal.2021.105495
15. Langelandsvik G, Horgar A, Furu T, et al. (2020) Comparative study of eutectic Al-Si alloys manufactured by WAAM and casting. *Int J Adv Manuf Technol* 110. https://doi.org/10.1007/s00170-020-05735-7
16. Langelandsvik G, Akselsen OM, Furu T, Roven HJ (2021) Review of aluminum alloy development for wire arc additive manufacturing. *Materials* 14: 5370.
17. McAndrew AR, Alvarez Rosales M, Colegrove PA, et al. (2018) Interpass rolling of Ti-6Al-4V wire + arc additively manufactured features for microstructural refinement. *Addit Manuf* 21. https://doi.org/10.1016/j.addma.2018.03.006

18. Lin Z, Song K, Yu X (2021) A review on wire and arc additive manufacturing of titanium alloy. *J Manuf Process* 70: 24–45. https://doi.org/10.1016/j.jmapro.2021.08.018
19. Dhinakaran V, Ajith J, Fathima Yasin Fahmidha A, et al. (2020) Wire arc additive manufacturing (WAAM) process of nickel based superalloys-A review. *Mater Today Proc* 21: 920–925.
20. Qiu Z, Wu B, Zhu H, et al. (2020) Microstructure and mechanical properties of wire arc additively manufactured Hastelloy C276 alloy. *Mater Des* 195. https://doi.org/10.1016/j.matdes.2020.109007
21. Hassel T, Carstensen T (2020) Properties and anisotropy behaviour of a nickel base alloy material produced by robot-based wire and arc additive manufacturing. *Weld World* 64. https://doi.org/10.1007/s40194-020-00971-7
22. Baby J, Amirthalingam M (2020) Microstructural development during wire arc additive manufacturing of copper-based components. *Weld World* 64: 395–405. https://doi.org/10.1007/s40194-019-00840-y
23. Guo Y, Quan G, Celikin M, et al. (2022) Effect of heat treatment on the microstructure and mechanical properties of AZ80M magnesium alloy fabricated by wire arc additive manufacturing. *J Magnes Alloys* 10. https://doi.org/10.1016/j.jma.2021.04.006
24. Wang P, Zhang H, Zhu H, et al. (2021) Wire-arc additive manufacturing of AZ31 magnesium alloy fabricated by cold metal transfer heat source: Processing, microstructure, and mechanical behavior. *J Mater Process Technol* 288. https://doi.org/10.1016/j.jmatprotec.2020.116895
25. Cobb HM (2008) The naming and numbering of stainless steels. *Iron Steel Technol* 5. https://doi.org/10.31399/asm.tb.hss.t52790241
26. Zhang S, Wang Q, Yang R, Dong C (2021) Composition equivalents of stainless steels understood via gamma stabilizing efficiency. *Sci Rep* 11: 5423. https://doi.org/10.1038/s41598-021-84917-z
27. Sales A, Kotousov A, Perilli E, Yin L (2022) Improvement of the fatigue resistance of super duplex stainless-steel (SDSS) components fabricated by wire arc additive manufacturing (WAAM). *Metals (Basel)* 12. https://doi.org/10.3390/met12091548
28. Dasarathy C (1993) Steels – metallurgy and applications. *Mater Des* 14. https://doi.org/10.1016/0261-3069(93)90100-a
29. Moore P (2013) *Technical Handbook of Stainless Steels*. Atlas Steels.
30. Derekar KS (2018) A review of wire arc additive manufacturing and advances in wire arc additive manufacturing of aluminium. *Mater Sci Technol (United Kingdom)* 34: 895–916. https://doi.org/10.1080/02670836.2018.1455012
31. Arvieu C, Galy C, Le Guen E, Lacoste E (2020) Relative density of SLM-produced aluminum alloy parts: Interpretation of results. *J Manuf Mater Proces* 4. https://doi.org/10.3390/JMMP4030083
32. Langelandsvik G, Akselsen OM, Furu T, Roven HJ (2021) Review of aluminum alloy development for wire arc additive manufacturing. *Materials* 14: 5370.
33. Takata N, Liu M, Kodaira H, et al. (2020) Anomalous strengthening by supersaturated solid solutions of selectively laser melted Al–Si-based alloys. *Addit Manuf* 33. https://doi.org/10.1016/j.addma.2020.101152
34. Yilmaz A (2011) The Portevin-Le Chatelier effect: A review of experimental findings. *Sci Technol Adv Mater* 12:6: https://doi.org/10.1088/1468-6996/12/6/063001
35. Uddin SZ, Murr LE, Terrazas CA, et al. (2018) Processing and characterization of crack-free aluminum 6061 using high-temperature heating in laser powder bed fusion additive manufacturing. *Addit Manuf* 22. https://doi.org/10.1016/j.addma.2018.05.047
36. Omiyale BO, Olugbade TO, Abioye TE, Farayibi PK (2022) Wire arc additive manufacturing of aluminium alloys for aerospace and automotive applications: A review. *Mater Sci Technol (United Kingdom)* 38: 391–408.

37. Thapliyal S (2019) Challenges associated with the wire arc additive manufacturing (WAAM) of aluminum alloys. *Mater Res Express* 6: 112006.
38. Gu J, Yang S, Gao M, et al. (2020) Micropore evolution in additively manufactured aluminum alloys under heat treatment and inter-layer rolling. *Mater Des* 186. https://doi.org/10.1016/j.matdes.2019.108288
39. Dhinakaran V, Ajith J, Fathima Yasin Fahmidha A, et al. (2020) Wire arc additive manufacturing (WAAM) process of nickel based superalloys-A review. *Mater Today Proc* 21: 920–925.
40. Senthil TS, Ramesh Babu S, Puviyarasan M, Dhinakaran V (2021) Mechanical and microstructural characterization of functionally graded Inconel 825 – SS316L fabricated using wire arc additive manufacturing. *J Mater Res Technol* 15. https://doi.org/10.1016/j.jmrt.2021.08.060
41. Paul S, Liu J, Strayer ST, et al. (2020) A discrete dendrite dynamics model for epitaxial columnar grain growth in metal additive manufacturing with application to inconel. *Addit Manuf* 36. https://doi.org/10.1016/j.addma.2020.101611
42. Schroepfer D, Treutler K, Boerner A, et al. (2021) Surface finishing of hard-to-machine cladding alloys for highly stressed components. *Int J Adv Manuf Technol* 114. https://doi.org/10.1007/s00170-021-06815-y
43. Mumtaz KA, Erasenthiran P, Hopkinson N (2008) High density selective laser melting of Waspaloy®. *J Mater Process Technol* 195. https://doi.org/10.1016/j.jmatprotec.2007.04.117
44. Qiu Z, Dong B, Wu B, et al. (2021) Tailoring the surface finish, dendritic microstructure and mechanical properties of wire arc additively manufactured Hastelloy C276 alloy by magnetic arc oscillation. *Addit Manuf* 48. https://doi.org/10.1016/j.addma.2021.102397
45. Qiu Z, Wu B, Wang Z, et al. (2021) Effects of post heat treatment on the microstructure and mechanical properties of wire arc additively manufactured Hastelloy C276 alloy. *Mater Charact* 177. https://doi.org/10.1016/j.matchar.2021.111158
46. Dinovitzer M, Chen X, Laliberte J, et al. (2019) Effect of wire and arc additive manufacturing (WAAM) process parameters on bead geometry and microstructure. *Addit Manuf* 26. https://doi.org/10.1016/j.addma.2018.12.013
47. Bhanu V, Fydrych D, Gupta A, Pandey C (2021) Study on microstructure and mechanical properties of laser welded dissimilar joint of p91 steel and incoloy 800ht nickel alloy. *Materials* 14. https://doi.org/10.3390/ma14195876
48. Rogalski G, Świerczyńska A, Landowski M, Fydrych D (2020) Mechanical and microstructural characterization of tig welded dissimilar joints between 304l austenitic stainless steel and incoloy 800ht nickel alloy. *Metals (Basel)* 10. https://doi.org/10.3390/met10050559
49. Raja M, Tiwari Y, Mukherjee M, et al. (2022) Effect of bidirectional and switchback deposition strategies on microstructure and mechanical properties of wire arc additive manufactured Inconel 625. *Int J Adv Manuf Technol* 119: 4845–4861. https://doi.org/10.1007/s00170-022-08687-2
50. Guo C, Ying M, Dang H, et al. (2021) Microstructural and intergranular corrosion properties of Inconel 625 superalloys fabricated using wire arc additive manufacturing. *Mater Res Express* 8. https://doi.org/10.1088/2053-1591/abe977
51. Feng Z, Wen Z, Lu G, Zhao Y (2022) Influence of cooling scenarios on the evolution of microstructures in nickel-based single crystal superalloys. *Crystals (Basel)* 12. https://doi.org/10.3390/cryst12010074
52. Mignanelli PM, Jones NG, Pickering EJ, Messé OMDM, Rae CMF, Hardy MC, et al. (2017) Gamma-gamma prime-gamma double prime dual-superlattice superalloys. *Scr Mater* 136: 136–140. https://doi.org/10.1016/j.scriptamat.2017.04.029

53. Tiwari Y, Mukherjee M, Chatterjee D, Manivannan R (2023) Influence of inter-layer rotation in parallel deposition strategies on the microstructure, texture, and mechanical behaviour of Inconel-625 during directed energy deposition. *Mater Charact* 197. https://doi.org/10.1016/j.matchar.2023.112711
54. Lin Z, Song K, Yu X (2021) A review on wire and arc additive manufacturing of titanium alloy. *J Manuf Process* 70: 24–45.
55. Polmear I, St. John D, Nie JF, Qian M (2017) *Light Alloys: Metallurgy of the Light Metals*. 5th Ed., Butterworth-Heinemann
56. Geetha M, Singh AK, Asokamani R, Gogia AK (2009) Ti based biomaterials, the ultimate choice for orthopaedic implants – A review. *Prog Mater Sci* 54: 397–425.
57. Askeland DR (1994) The Science and Engineering of Materials. *Eur J Eng Educ* 19. https://doi.org/10.1080/03043799408928327
58. Dutta Majumdar J, Manna I (2015) Laser Surface Engineering of Titanium and Its Alloys for Improved Wear, Corrosion and High-Temperature Oxidation Resistance. In: *Laser Surface Engineering: Processes and Applications*, J. Lawrence, D.G. Waugh (Eds), Woodhead publicatios, pp. 483–521.
59. Bassani F, Liedl GL, Wyder P (2005) *Encyclopedia of condensed matter physics*. Academic Press.
60. Ahmed T, Rack HJ (1998) Phase transformations during cooling in $\alpha + \beta$ titanium alloys. *Mater Sci Eng A* 243. https://doi.org/10.1016/s0921-5093(97)00802-2
61. Donachie MJ (2000) *Titanium – A Technical Guide*. 2nd Ed., 55, ASM International
62. Gangwar K, Ramulu M (2018) Friction stir welding of titanium alloys: A review. *Mater Des* 141. https://doi.org/10.1016/j.matdes.2017.12.033
63. Okamoto H (2010) *Phase Diagrams for Binary Alloys*. ASM International.
64. Massalski T, Subramanian P (2016) Hf (Hafnium) Binary Alloy Phase Diagrams. In: *Alloy Phase Diagrams*, H. Okamoto; M.E. Schlesinger; E.M. Mueller (Eds). ASM international.
65. Jeje SO, Shongwe MB, Rominiyi AL, Olubambi PA (2021) Spark plasma sintering of titanium matrix composite—A review. *Int J Adv Manuf Technol* 117.
66. Ralph B (2008) Titanium alloys: An atlas of structures and fracture features. *Mater Charact* 59. https://doi.org/10.1016/j.matchar.2007.01.007
67. Li H, Zhao Z, Ning Y, et al. (2018) Characterization of microstructural evolution for a near-α Titanium alloy with different initial lamellar microstructures. *Metals (Basel)* 8. https://doi.org/10.3390/met8121045
68. Tan C, Fan Y, Sun Q, Zhang G (2020) Improvement of the crack propagation resistance in an $\alpha + \beta$ titanium alloy with a trimodal microstructure. *Metals (Basel)* 10: 1–11. https://doi.org/10.3390/met10081058
69. Kolli RP, Devaraj A (2018) A review of metastable beta titanium alloys. *Metals (Basel)* 8: 506.
70. Marinelli G, Martina F, Ganguly S, Williams S (2019) Development of wire + arc additive manufacture for the production of large-scale unalloyed tungsten components. *Int J Refract Met Hard Mater* 82: 329–335. https://doi.org/10.1016/j.ijrmhm.2019.05.009
71. Marinelli G, Martina F, Ganguly S, et al. (2019) Microstructure and thermal properties of unalloyed tungsten deposited by Wire + Arc Additive Manufacture. *J Nucl Mater* 522: 45–53. https://doi.org/10.1016/j.jnucmat.2019.04.049
72. Marinelli G, Martina F, Lewtas H, et al. (2019) Functionally graded structures of refractory metals by wire arc additive manufacturing. *Sci Technol Weld Join* 24: 495–503. https://doi.org/10.1080/13621718.2019.1586162
73. Köck W, Paschen P (1989) Tantalum-processing, properties and applications. *JOM* 41. https://doi.org/10.1007/BF03220360

74. Marinelli G, Martina F, Ganguly S, Williams S (2020) Grain refinement in an unalloyed tantalum structure by combining Wire+Arc additive manufacturing and vertical cold rolling. *Addit Manuf* 32. https://doi.org/10.1016/j.addma.2019.101009
75. Marinelli G, Martina F, Ganguly S, Williams S (2019) Microstructure, hardness and mechanical properties of two different unalloyed tantalum wires deposited via wire + arc additive manufacture. *Int J Refract Met Hard Mater* 83. https://doi.org/10.1016/j.ijrmhm.2019.104974
76. Bechtold JH (1955) Tensile properties of annealed tantalum at low temperatures. *Acta Metall* 3. https://doi.org/10.1016/0001-6160(55)90060-2
77. Martina F, Colegrove PA, Williams SW, Meyer J (2015) Microstructure of interpass rolled wire + arc additive manufacturing Ti-6Al-4V components. *Metall Mater Trans A Phys Metall Mater Sci* 46. https://doi.org/10.1007/s11661-015-3172-1
78. Maury N, Zhang NX, Huang Y, et al. (2015) A critical examination of pure tantalum processed by high-pressure torsion. *Mater Sci Eng A* 638. https://doi.org/10.1016/j.msea.2015.04.053
79. Mathaudhu SN, Ted Hartwig K (2006) Grain refinement and recrystallization of heavily worked tantalum. *Mater Sci Eng A* 426. https://doi.org/10.1016/j.msea.2006.03.089
80. Aditya AV, Subramanian PK, Gopala Krishna V, Chinta Babu U (2012) Influence of rolling path on microstructure and mechanical properties in EB refined tantalum. *Transac Indian Inst Metals* 65. https://doi.org/10.1007/s12666-012-0148-3
81. El-Genk MS, Tournier JM (2005) A review of refractory metal alloys and mechanically alloyed-oxide dispersion strengthened steels for space nuclear power systems. *J Nucl Mater* 340: 93–112.
82. Voronova LM, Chashchukhina TI, Degtyarev M V. (2018) Recrystallization Texture of Submicrocrystalline Niobium after Annealing. *Phys Met Metallogr* 119. https://doi.org/10.1134/S0031918X18090156
83. Tang HP, Yang GY, Jia WP, et al. (2015) Additive manufacturing of a high niobium-containing titanium aluminide alloy by selective electron beam melting. *Mater Sci Eng A* 636. https://doi.org/10.1016/j.msea.2015.03.079
84. Luo C, Wang C, Wu X, et al. (2021) Influence of porous tantalum scaffold pore size on osteogenesis and osteointegration: A comprehensive study based on 3D-printing technology. *Mater Sci Eng C* 129. https://doi.org/10.1016/j.msec.2021.112382
85. Liu Y, Bao C, Wismeijer D, Wu G (2015) The physicochemical/biological properties of porous tantalum and the potential surface modification techniques to improve its clinical application in dental implantology. *Mater Sci Eng C* 49: 323–329.
86. Lee J, Park SY, Jeong H (2020) Effect of strains on textural evolution of hydrostatically extruded niobium tubes. *Mater Sci Technol (United Kingdom)* 36. https://doi.org/10.1080/02670836.2020.1759191
87. Yang PJ, Li QJ, Tsuru T, et al. (2019) Mechanism of hardening and damage initiation in oxygen embrittlement of body-centred-cubic niobium. *Acta Mater* 168. https://doi.org/10.1016/j.actamat.2019.02.030
88. Liu G, Zhang X, Chen X, et al. (2021) Additive manufacturing of structural materials. *Mater Sci Eng R Rep* 145: 100596.
89. Tapia G, Elwany A (2014) A review on process monitoring and control in metal-based additive manufacturing. *J Manuf Sci E T ASME* 136: 060801.
90. Mazey DJ, English CA (1984) Role of refractory metal alloys in fusion reactor applications. *J Less-Common Met* 100. https://doi.org/10.1016/0022-5088(84)90078-X
91. Wang J, Liu C, Lu T, et al. (2022) Microstructure and mechanical properties of unalloyed molybdenum fabricated via wire arc additive manufacturing. *Int J Refract Met Hard Mater* 107. https://doi.org/10.1016/j.ijrmhm.2022.105886

92. Martin JH, Yahata BD, Hundley JM, et al. (2017) 3D printing of high-strength aluminium alloys. *Nature* 549. https://doi.org/10.1038/nature23894
93. Xu T, Tang S, Liu C, et al. (2020) Obtaining large-size pyramidal lattice cell structures by pulse wire arc additive manufacturing. *Mater Des* 187. https://doi.org/10.1016/j.matdes.2019.108401
94. Kaserer L, Braun J, Stajkovic J, et al. (2019) Fully dense and crack free molybdenum manufactured by Selective Laser Melting through alloying with carbon. *Int J Refract Met Hard Mater* 84. https://doi.org/10.1016/j.ijrmhm.2019.105000
95. Faidel D, Jonas D, Natour G, Behr W (2015) Investigation of the selective laser melting process with molybdenum powder. *Addit Manuf* 8. https://doi.org/10.1016/j.addma.2015.09.002
96. Wang D, Yu C, Ma J, et al. (2017) Densification and crack suppression in selective laser melting of pure molybdenum. *Mater Des* 129. https://doi.org/10.1016/j.matdes.2017.04.094
97. Braun J, Kaserer L, Stajkovic J, et al. (2019) Molybdenum and tungsten manufactured by selective laser melting: Analysis of defect structure and solidification mechanisms. *Int J Refract Met Hard Mater* 84. https://doi.org/10.1016/j.ijrmhm.2019.104999
98. Bhardwaj T, Shukla M, Paul CP, Bindra KS (2019) Direct energy deposition – Laser additive manufacturing of titanium-molybdenum alloy: Parametric studies, microstructure and mechanical properties. *J Alloys Compd* 787. https://doi.org/10.1016/j.jallcom.2019.02.121
99. Higashi M, Ozaki T (2020) Selective laser melting of pure molybdenum: Evolution of defect and crystallographic texture with process parameters. *Mater Des* 191. https://doi.org/10.1016/j.matdes.2020.108588
100. Wang J, Cui Y, Liu C, et al. (2020) Understanding internal defects in Mo fabricated by wire arc additive manufacturing through 3D computed tomography. *J Alloys Compd* 840. https://doi.org/10.1016/j.jallcom.2020.155753
101. DebRoy T, Wei HL, Zuback JS, et al. (2018) Additive manufacturing of metallic components – Process, structure and properties. *Prog Mater Sci* 92: 112–224.
102. Qi J, Wang J, Xu T, Liu C (2023) Analysis of external and internal defects of molybdenum deposited via wire arc additive manufacturing. *Mater Lett* 336. https://doi.org/10.1016/j.matlet.2023.133880
103. Orzolek SM, Semple JK, Fisher CR (2022) Influence of processing on the microstructure of nickel aluminum bronze (NAB). *Addit Manuf* 56: 102859.
104. Kim J, Kim J, Pyo C (2020) Comparison of mechanical properties of ni-al-bronze alloy fabricated through wire arc additive manufacturing with ni-al-bronze alloy fabricated through casting. *Metals (Basel)* 10: 1–15. https://doi.org/10.3390/met10091164
105. Cai X, Yang M, Wang S, et al. (2023) Experimental investigations on corrosion behavior and antibacterial property of nickel-aluminum bronze fabricated through wire-arc additive manufacturing (WAAM). *Corros Sci* 214. https://doi.org/10.1016/j.corsci.2023.111040
106. Dharmendra C, Amirkhiz BS, Lloyd A, et al. (2020) Wire-arc additive manufactured nickel aluminum bronze with enhanced mechanical properties using heat treatments cycles. *Addit Manuf* 36. https://doi.org/10.1016/j.addma.2020.101510
107. Ding D, Pan Z, van Duin S, et al. (2016) Fabricating superior NiAl bronze components through wire arc additive manufacturing. *Materials* 9. https://doi.org/10.3390/ma9080652
108. Shen C, Hua X, Reid M, et al. (2020) Thermal induced phase evolution of Fe–Fe3Ni functionally graded material fabricated using the wire-arc additive manufacturing process: An in-situ neutron diffraction study. *J Alloys Compd* 826. https://doi.org/10.1016/j.jallcom.2020.154097

109. Wang J, Pan Z, Ma Y, et al. (2018) Characterization of wire arc additively manufactured titanium aluminide functionally graded material: Microstructure, mechanical properties and oxidation behaviour. *Mater Sci Eng A* 734: 110–119. https://doi.org/10.1016/j.msea.2018.07.097
110. Wang J, Pan Z, Carpenter K, et al. (2021) Comparative study on crystallographic orientation, precipitation, phase transformation and mechanical response of Ni-rich NiTi alloy fabricated by WAAM at elevated substrate heating temperatures. *Mater Sci Eng A* 800. https://doi.org/10.1016/j.msea.2020.140307
111. Wang J, Pan Z, Yang G, et al. (2019) Location dependence of microstructure, phase transformation temperature and mechanical properties on Ni-rich NiTi alloy fabricated by wire arc additive manufacturing. *Mater Sci Eng A* 749: 218–222. https://doi.org/10.1016/j.msea.2019.02.029
112. Resnina N, Palani IA, Belyaev S, et al. (2021) Structure, martensitic transformations and mechanical behaviour of NiTi shape memory alloy produced by wire arc additive manufacturing. *J Alloys Compd* 851. https://doi.org/10.1016/j.jallcom.2020.156851
113. Zeng Z, Cong BQ, Oliveira JP, et al. (2020) Wire and arc additive manufacturing of a Ni-rich NiTi shape memory alloy: Microstructure and mechanical properties. *Addit Manuf* 32. https://doi.org/10.1016/j.addma.2020.101051
114. Zeng Z, Cong BQ, Oliveira JP, et al. (2020) Wire and arc additive manufacturing of a Ni-rich NiTi shape memory alloy: Microstructure and mechanical properties. *Addit Manuf* 32. https://doi.org/10.1016/j.addma.2020.101051
115. Chen X, Yu G, He X, et al. (2018) Effect of droplet impact on molten pool dynamics in hybrid laser-MIG welding of aluminum alloy. *Int J Adv Manuf Technol* 96: 209–222. https://doi.org/10.1007/s00170-017-1509-1
116. Zhang Y, Wu S, Cheng F (2022) A specially-designed super duplex stainless steel with balanced ferrite:austenite ratio fabricated via flux-cored wire arc additive manufacturing: Microstructure evolution, mechanical properties and corrosion resistance. *Mater Sci Eng A* 854. https://doi.org/10.1016/j.msea.2022.143809
117. Zhang Y, Cheng F, Wu S (2021) The microstructure and mechanical properties of duplex stainless steel components fabricated via flux-cored wire arc-additive manufacturing. *J Manuf Process* 69: 204–214. https://doi.org/10.1016/j.jmapro.2021.07.045
118. Wang Z, Bai X, Que M, Zhou X (2023) Wire arc additive manufacturing of network microstructure (TiB+TiC)/Ti6Al4V composites using flux-cored wires. *Ceram Int* 49: 4168–4176. https://doi.org/10.1016/j.ceramint.2022.09.299
119. Treutler K, Lorenz S, Hamje J, Wesling V (2022) Wire and Arc Additive Manufacturing of a CoCrFeMoNiV Complex Concentrated Alloy Using Metal-Cored Wire—Process, Properties, and Wear Resistance. *Appl Sci (Switzerland)* 12. https://doi.org/10.3390/app12136308
120. Lin Z, Goulas C, Ya W, Hermans MJM (2019) Microstructure and mechanical properties of medium carbon steel deposits obtained via wire and arc additive manufacturing using metal-cored wire. *Metals (Basel)* 9. https://doi.org/10.3390/met9060673
121. Chaudhari R, Parikh N, Khanna S, et al. (2022) Effect of multi-walled structure on microstructure and mechanical properties of 1.25Cr-1.0Mo steel fabricated by GMAW-based WAAM using metal-cored wire. *J Mater Res Technol* 21: 3386–3396. https://doi.org/10.1016/j.jmrt.2022.10.158
122. Mvola B, Kah P (2017) Effects of shielding gas control: Welded joint properties in GMAW process optimization. *Int J Adv Manuf Technol* 88: 2369–2387.
123. Kah P, Martikainen J (2013) Influence of shielding gases in the welding of metals. *Int J Adv Manuf Technol* 64: 1411–1421. https://doi.org/10.1007/s00170-012-4111-6

7 Wire Arc Additive Manufacturing of Non-ferrous Alloys

Basant Kumar, Sheikh Nazir Ahmad, and Sandeep Rathee
National Institute of Technology Srinagar, Jammu & Kashmir, India

Manu Srivastava
PDPM Indian Institute of Information Technology, Design, and Manufacturing, Jabalpur, India

7.1 INTRODUCTION TO ADDITIVE MANUFACTURING (AM)

AM is a group of technologies that enables the fabrication of objects by adding the material to a substrate or pre-build layer. It is a fast-evolving technology widely used by manufacturing organisations thanks to the enhancing ability of this technique for building large products with intricate designs at relatively high deposition rates [1]. The definition of AM as per the American Society for Testing and Materials ASTM standards is: "AM is a layer-by-layer process of depositing materials to produce objects from 3D model data. This contrasts with subtractive manufacturing techniques, which call for the removal of material to achieve the desired shape or form" [2, 3]. Figure 7.1 represents the comparison between AM and subtractive manufacturing. As per ASTM standards, AM techniques are divided into seven categories: directed energy deposition technique, binder jetting, powder bed fusion (PBF), material extrusion, material jetting, vat polymerisation, and sheet lamination [4]. In PBF technology, the material is melted across a surface to build components using a laser or electron beam [5].

During this procedure, the platform is covered with a layer of material before being joined by a laser or electron beam. After the first layer has been fused, a second layer is added on top of it using a roller, and the energy source is then used to fuse it once more. This procedure is repeated until the complete part has been produced. The produced component has better precision since the raw material is in powder form, which enables it to construct more complex geometries; nevertheless, these processes are time-consuming for making large components [6]. Aluminum foils or paper-based filaments are cut with a very sharp blade or laser during the sheet lamination process, and they are then glued together [7]. The binder jetting procedure

DOI: 10.1201/9781003363415-7

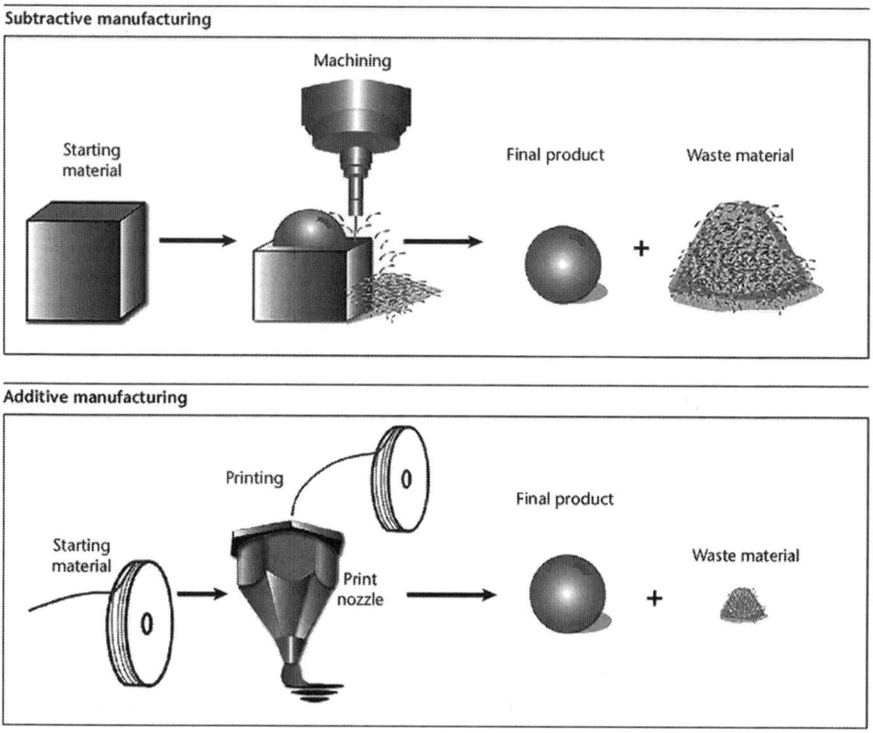

FIGURE 7.1 Subtractive and additive manufacturing [31].

makes use of a portion of the powder bed that is specifically joined together using a liquid bonding agent, such as the PBF procedure [8]. In DED, the material is melted with the aid of a nozzle as it is deposited to create pieces. The extant AM literature lists a few drawbacks of the available DED systems, such as the need for vacuum in the electron-based system compared to the laser-based system, and due to the melting of less metal powder, metal wire deposition has a higher deposition efficiency compared to metal powder deposition [9, 10].

AM will significantly affect the manufacturing sector of the future which has been anticipated by several experts [11]. AM is gaining popularity because of its main advantages, including its ability to work with a wide range of materials, including ceramics, polymers, and metals, as well as its capacity to make innovative, complicated, and almost net-shaped components that do not require further tooling or re-fixturing. AM has the potential to lower overall manufacturing costs by utilising a specialised manufacturing process that significantly lowers the lead time and material wastage while enhancing feedback versatility for converting feedstock into a desired structure [12]. Figure 7.2 depicts the different varieties of AM with their material compatibility. DED is a part of AM, and WAAM is identified as a part of DED. This chapter substantially focuses on WAAM in the subsequent sections.

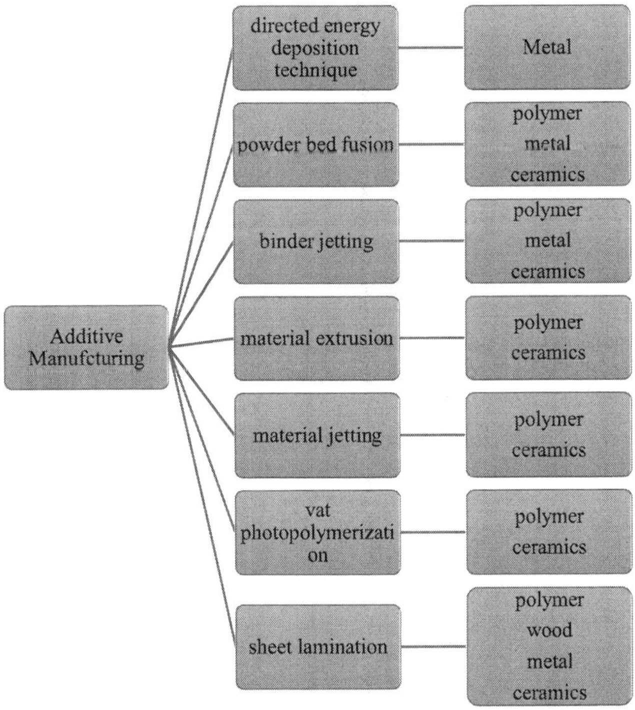

FIGURE 7.2 Classification of AM with material compatibility.

7.2 WIRE ARC ADDITIVE MANUFACTURING (WAAM)

The limitations of laser and electron-beam-based AM processes for powder and wire are overcome by WAAM. It offers several advantages, including minimal material waste, cost-effectiveness, flexibility in equipment usage, rapid production cycles, and excellent forming quality [13–16]. The metal wire is used as the feedstock for WAAM, which is categorised as a DED process by ISO/ASTM 52900 [17]. The wire is precisely fed into the welding arc using a wire feeder, where it completely melts and gets deposited on an earlier applied layer or substrate, a portion of which also melts in the process [17, 18]. With a 0.2 mm dimensional accuracy, fully dense components can be produced using the WAAM technique [19, 20]. Unlike conventional casting techniques, WAAM enables the production of components with thin or thick walls, solid structures, or cast-like geometries, all while obtaining enhanced mechanical properties and a finer microstructure without the use of dies [20, 21]. High deposition rates across multiple welding processes are another benefit of WAAM. For instance, the rate of Ti6Al4V deposition utilising laser PBF was 0.1 to 0.18 kg/h [22], but the rate for electron beam PBF was 0.26 to 0.36 kg/h [23]. WAAM, on the other hand, may produce deposition rates of 0.5 to 4 kg/h [24]. Furthermore, WAAM has outstanding material utilisation efficiency, achieving rates of up to 90% [20]. Even though some final machining may still be necessary, when compared to PBF procedures, it is often reduced to about 50% [5]. WAAM has become a potential process for a variety of materials, including titanium, aluminium, magnesium, and steel, as a result of its considerable advantages. As an alternative to traditional manufacturing processes that rely on raw material processing, this technology is increasingly being utilised [25]. Using WAAM widely is intended to build items by depositing materials in layers. However, much like every manufacturing procedure, WAAM has several issues that must be resolved to improve the calibre of the components generated. These main WAAM roadblocks are also explored in the present chapter, which also covers several methods for improving the quality of WAAMed components. It also emphasises the research prospects in the WAAM domain as it comes to a conclusion.

7.3 CLASSIFICATIONS OF WAAM BASED ON WELDING TECHNIQUES

The welding technology utilising arc as a heat source for melting is being used in the WAAM process, based upon which the WAAM process can be classified into three categories: methods based on gas tungsten arc welding (GTAW), gas metal arc welding (GMAW) and plasma arc welding (PAW). WAAM welding processes are chosen depending on the feedstock material and application.

The GMAW technique makes use of a consumable wire electrode, continuous feed, and a power source to produce an electrical arc between the electrode and the workpiece. This electrical arc heats the metals, which in turn allows them to join together. In addition, a shielding gas (argon) is utilised so that the weld does not become contaminated by its surrounding environment. In GMAW, the material is deposited by holding the torch at an appropriate angle to the workpiece. The rate of molten material deposition is 1–2 kg/hour [26, 27]. The metal transfer modes onto the substrate of the GMAW process include globular, short-circuit spray, and pulsed spray. The surface finish and the wall thickness of the product made are among the primary limitations of the GMAW method [28–30]. Figure 7.3 represents the schematic of GMAW-based WAAM.

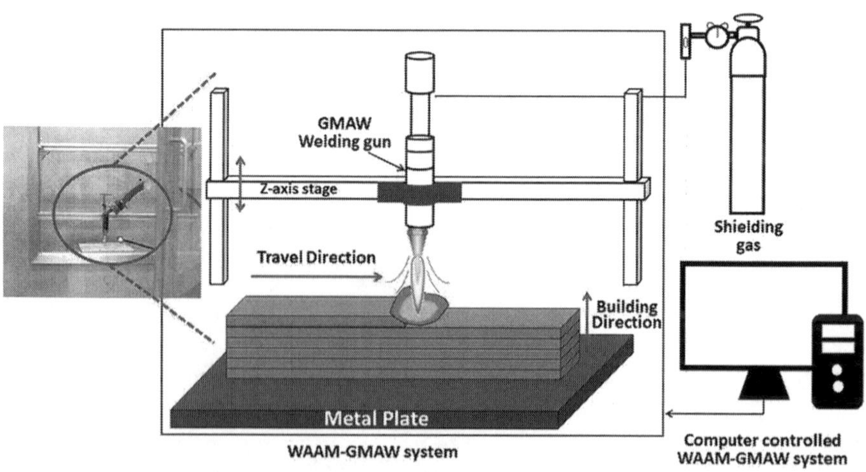

FIGURE 7.3 Schematic of GMAW-based WAAM [32].

Tungsten inert gas (TIG) Welding, more often known as GTAW, is a method of arc welding that generates a weld by employing a tungsten electrode that does not deplete during the course of the operation. During the process of TIG welding, an extremely intense electric arc is created between a tungsten electrode that is practically non-consumable and the piece of metal that is being welded. This arc supplies the heat that is essential for melting the metal, which is followed by the process of joining the pieces together. During the melting phase of the welding process, the electrode does not melt and becomes a part of the weld being created. Following a similar procedure to that used in oxyacetylene welding, a welding rod is introduced into the weld zone alongside the base metal when welding joints that necessitate filler metal. This is similar to how oxyacetylene welding is performed. An inert gas is ducted straight to the weld zone, where it surrounds the tungsten and acts as a barrier between it and the environment. This gas provides protection for the weld zone from the air that is outside the weld zone. Argon and helium are the two examples of inert gases that are utilised most frequently [33, 34]. In the case of GTAW, the rate of deposition is between 3 and 4 kg per hour. The GTAW has a few shortcomings, including high levels of welding spatter and poor arc stability. Figure 7.4 shows a diagrammatic representation of the GTAW technique. Feed wire for welding and depositing material are separate in the case of the GTAW technique.

Welding is done with a plasma jet in PAW. Plasma is a highly ionised gas that contains both positive and negative ions, has a high temperature, and conducts current readily. Argon gas and a tungsten electrode are utilised in PAW. The plasma jet is created with argon gas, which also serves as a shielding gas. When an arc is fed through a constructive nozzle, a plasma jet is formed. As a result, the plasma jet will adopt a narrow columnar structure with distinct features, making it perfect for welding. The deposition rate is typically in the range of 2–4 kg/hour in the case of PAW. Figure 7.5 represents the systematic diagram of the PAW technique.

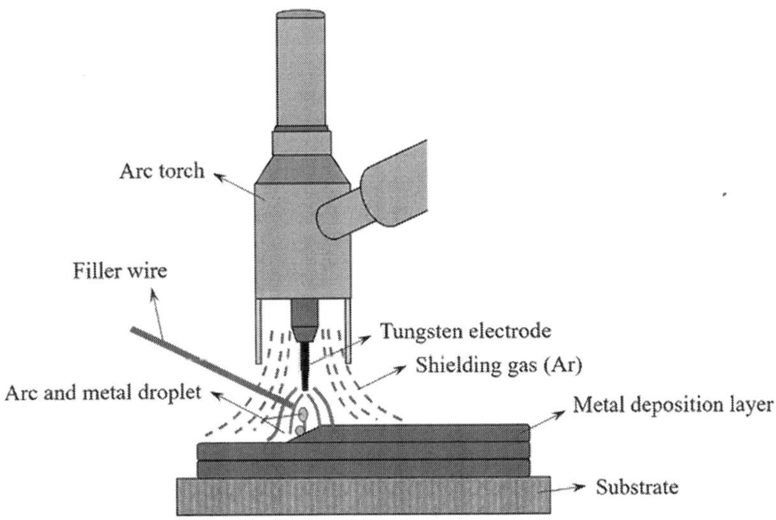

FIGURE 7.4 Schematic of GTAW-based WAAM [35].

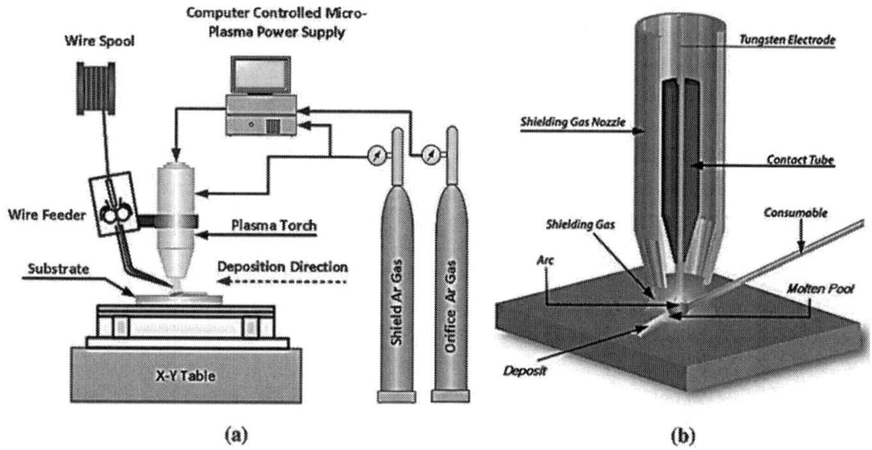

FIGURE 7.5 Schematic diagram of the PAW technique [38].

WAAM technique based on the GMAW procedure has practically two to three times increased deposition rate than the GTAW and PAW-based WAAM techniques when evaluated among the three welding technologies that are applied in the WAAM process [36]. In GTAW and PAW, unlike GMAW, the direction of the wire feed can change, which has an effect on the uniformity of the deposit and makes process planning more problematic [8]. GMAW has the highest deposition rate and is the most cost-effective of the WAAM welding processes, although it does have significant issues that need to be addressed. Heat input and weld spatter are both on the high side, resulting in lower weld quality and, as a result, lower WAAM-manufactured component quality [25]. To overcome these flaws, cold metal transfer (CMT) came into the picture. CMT includes a consumable wire electrode, which is reciprocating in nature. Weld quality was increased with less heat input and spatter disturbances. Normal, pulse, advanced, and advanced pulse are the many types of metal transfer modes that can be utilised when doing CMT welding [37].

7.4 BASIC COMPONENTS OF WAAM

In WAAM processes, layers of weld metal beads are deposited through a welding system in conjunction with a tool to bring the welding torch to the required position, such as a robot or a computer numerical control (CNC) milling machine, to create 3D metallic components. The welding deposition experimental setup consists of two separate systems: a welding system and a positioning system.

7.4.1 Welding System

GMAW, GTAW, and PAW methods may be used. The main requirement of the welding system is the adjustment of current intensity and voltage during deposition, which results in lower welding temperature and wire travel optimisation, as in CMT and DC pulsed current welding.

7.4.2 Positioning System

The system's general movement should be easy to control. The welding flame may be positioned on the z-axis by modifying a CNC milling machine. The welding torch is attached to the milling head at a defined z-level, and the table of the CNC system allows movement for the deposition of a welding layer. For the next layer deposition, the z-axis raises the torch, allowing the x-y table to move. The second positioning system that can position welding flame is a robotic arm.

7.5 MODES OF METAL TRANSFER

Depending on the metal being used, various types of metal transfer techniques are utilised. These have been discussed in this section. Each of these techniques has a unique deposition rate that is adapted to the unique properties of the metal involved.

7.5.1 Globular Metal Transfer

In comparison to short-circuit transfer, the welding current is set at a lower level, and the arc gap is purposefully made wider. This enables a slower, more gradual increase in droplet size. The droplet keeps expanding until gravity force is greater than surface tension. The transfer process begins when the droplet's size exceeds the diameter of the electrode.

7.5.2 Spray Metal Transfer

Because of the larger welding current density compared to globular transfer, there is a higher melting rate and more pinch force. The size of the produced droplets is noticeably smaller. Additionally, the greater welding current raises the temperature, which reduces the surface tension force.

7.5.3 Short-Circuit Metal Transfer

In short-circuit metal transfer, the welding current is low; that's why the arc grows slowly, and after the drop starts touching the weld pool, short-circuiting takes place. In this method, the molten drop doesn't attain a size that can make it fall under gravitational force. Welding current increases abruptly after starting to short-circuit, and excessive heat is generated. As a result, the molten metal is transferred to the weld pool, creating an arc gap that causes the arc voltage to suddenly increase. GMAW produces a short circuit when a liquid metal droplet attached to the electrode tip continuously makes contact with the weld pool. When the pool and electrode's liquid contact are broken, a new arc period starts. Then, magnetic forces and surface tension play a major role in moving molten metal to the melt pool.

7.5.4 Pulse Spray Transfer

In order to facilitate optimal performance, the power supply is designed to run between a high spray transfer current and a low background current. The power

TABLE 7.1
Different Modes of Metal Transfer

Modes of Metal Transfer	Deposition Rate (Kg/hr)
Globular	1.8–3.2
Short-circuit or dip transfer	0.9–2.7
Spray	2.7–5.4
Rotating spray	6.3–13.5
Pulsed spray	0.9–2.7

supply enables efficient supercooling of the weld pool during the background cycle. One droplet is moved from the electrode to the weld pool during each cycle. This welding method's combination of low background current and higher energy compared to short-circuit transfer makes it ideal for thick-section, out-of-position welding. This approach results in a higher average current and improved side wall fusion (Table 7.1).

7.6 CMT

To enable drop-by-drop deposition of weld material, the CMT process, invented by Austrian company Fronius in 2004, involves working at a reduced welding current and retracting the weld wire during a short-circuit condition. This process differs from GMAW in the method of mechanical droplet detachment. The primary distinction is that in GMAW, the wire advances continuously into the weld pool, whereas in CMT, the wire retracts at instant current flows and breaks the arc. The droplet detaches from the filler and fuses with the still-molten metal again. The wire then moves forward to create another arc. CMT utilises a controlled method of material deposition by a sophisticated wire feed system at low thermal input.

7.7 MATERIALS USED IN WAAM

Commercially available, spooled welding wires in a variety of alloys are used as feedstock in WAAM processes. In WAAM processes, alloys like copper, aluminium, stainless steel, and carbon steel are frequently used. It is crucial to consider the unique qualities that each alloy possesses when selecting a feedstock material. In WAAM processes, alloyed carbon steel and stainless steel are frequently used. These alloys are preferred due to their exceptional strength and resistance to corrosion. Aluminium is a preferred alloy for WAAM processes because of its light weight and excellent thermal conductivity. Copper is also used in WAAM processes due to its superior electrical conductivity and heat resistance, and an inert gas is used to stop the molten material from oxidising. It is essential to comprehend the various process options and the underlying principles in order to produce a part that is structurally sound, error-free, and dependable. In order to produce high-quality parts with increased accuracy, the WAAM process makes use of a potent laser, an inert

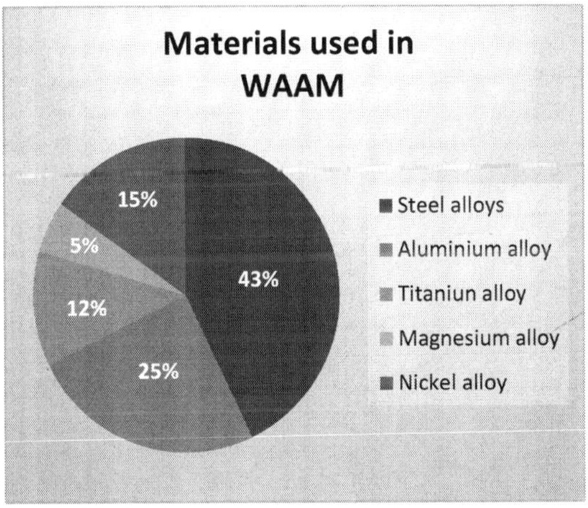

FIGURE 7.6 Chart depicting material used in WAAM.

gas, and a feedstock material. The laser melts the material, and an inert gas is used to keep the molten substance from oxidising. The feedstock material is solidified to create the part after being deposited in the desired shape. The WAAM process also permits the use of multiple layers of material to create more intricate parts. The final component is significantly influenced by both the WAAM process parameters and the chosen feedstock material. If the fundamentals of the WAAM procedure and the characteristics of the material are comprehended, parts can be fabricated optimally. The chart presented in Figure 7.6 illustrates the main WAAM materials.

7.8 WAAM OF NON-FERROUS ALLOYS

The demand for lighter components is rising in the automotive, marine, aviation, and aerospace industries because of environmental trends and the requirement to meet certificated emissions reduction (CER) standards. The trend of carbon trading is reducing utilisation of traditional parts manufacturing methods such as casting. The use of non-ferrous alloy components and AM techniques has significantly increased across a variety of industries because of the increasingly strict environmental regulations around the world. Meanwhile, reflective metal alloys with low laser coupling efficiency include Al and Mg alloys [39]. Furthermore, because of its high reactivity, powder metal is prone to generating a considerable fraction of porosity when used for AM of alloys having Mg and Al [40, 41]. Due to WAAM's 100% material and energy efficiency, it is independent of coupling efficiency. Consequently, WAAM of aluminum and magnesium alloys will be useful in the creation of high-value components for aviation, aerospace, biomedical, marine, and various other industries in the coming decade. Progress, use, challenges, and trends of non-ferrous WAAM techniques, especially with aluminum and magnesium alloys, are discussed in the subsequent subsections.

7.8.1 Nickel-Based Alloys

Alloys of Nickel are extensively being used in the petrochemical, marine, aerospace, and chemical industries because of their desirable incorporation of properties, including tensile strength, fatigue strength, and corrosion resistance in severe environments. Inconel 625 and Inconel 718 are two alloys of nickel which are frequently used in both research and industry. In this method, each layer goes through a unique heat cycle, which causes a variety of microstructures to emerge between the layers. Inconel 625 showed cellular grains in the lower part, columnar grains in the middle, and equiaxed grains in the upper region in research by J. F. Wang and colleagues [42]. Another research by Cheepu et al. [43] focused on building stacking systems for super-TIG welding to increase the microstructure of Inconel.

7.8.2 Titanium Alloys

Due to its excellent corrosion resistance, low density, and high strength-to-weight ratio, titanium alloy (Ti6Al4V) is a good choice for aerospace applications. When a WAAM item is machined from a forged billet, the buy-to-fly ratio may drop to 1.5 from its usual range of 10 to 20 [45]. There are two phases in Ti6Al4V alloy: a body-centred cubic structure and a hexagonal closed-packed structure [46]. In this technique, a temperature gradient is created because the deposition of material is done in layer-by-layer form, promoting the production of massive columnar grains [47]. It exhibits relatively low yield strength and ultimate strength, but it shows high elongation in the build direction, but opposite mechanical characteristics may be detected in the torch travel direction [48]. Figure 7.7 (a) and (b) illustrate a comparison of mechanical properties (yield strength/ultimate strength and elongation, respectively) for Ti6Al4V components manufactured with WAAM. This attribute is appropriate for components that require unidirectional characteristics.

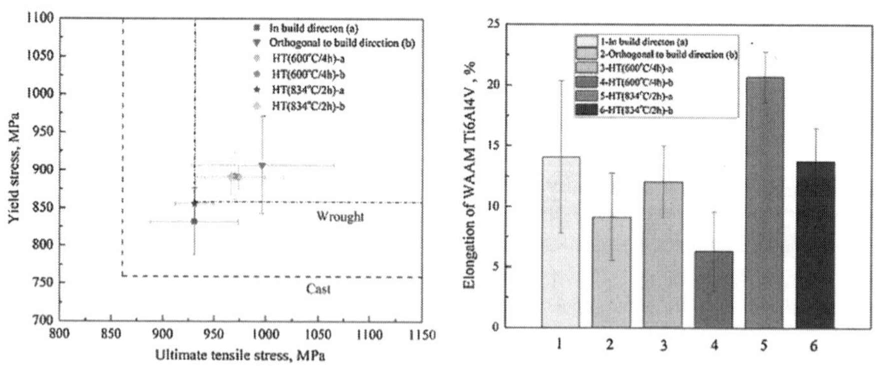

(a) (b)

FIGURE 7.7 Comparative analysis of the mechanical properties of Ti6Al4V components produced with WAAM: (a) tensile strength and (b) deformation of the material [36].

7.8.3 Aluminium Alloys

Porosity emerges as the main issue with WAAM for aluminium alloys, as per the recent research by Gu et al. [49]. Aluminium welding has historically been difficult, mostly because of various problems relating to the fundamental properties of the material. The formation of a sticking oxide layer, a high thermal expansion coefficient, a dual contraction during solidification in contrast to ferrous metals, and the presence of porosity are a few of these factors. Solidification cracking, which is closely related to the composition of the alloy and the amount of eutectic present during solidification, complicates aluminium welding. With higher alloy content, the susceptibility to cracking rises until it reaches a peak, especially in alloys like Al-Li, Al-Mg, Al-Cu, Al-Si, and Al-Mg-Si. Beyond this peak, however, the extra eutectic helps to fill the cracks and lessen crack sensitivity. While research on the use of filler metals from the 2xxx, 4xxx, and 5xxx series has been done, for the use of filler metal from the 7xxx series, there hasn't been much reported research.

7.8.4 Magnesium Alloys

Due to their wide range of impressive properties, magnesium alloys have recently received a lot of attention. These qualities include those that allow for welding, electromagnetic shielding, damping, biodegradability, and machining suitability [50]. Guo et al.'s study [51] revealed that the WAAM-fabricated AZ80M composition included the phases Mg, Mg17Al12, and Al2Y. The non-uniformity of the built direction caused the tensile strength to vary in both the vertical and horizontal directions. These differences were attributed to the microstructure's inconsistency and the presence of defects. Martinez et al.'s study [52] looked into the WAAM production of the AZ91D magnesium alloy, which contains 9% Al and 1% Zn. The findings showed that only the individual printed lines' surface showed signs of porosity or cracking. The magnesium matrix in the deposited centre portions, in contrast, had a hexagonal close-packed (HCP) structure. These areas were devoid of any voids and had a dense distribution of precipitates that were enriched in aluminium and zinc. Magnesium AZ61A alloy specimens were created using CMT in a different study by Klein et al. [53]. The researchers established that the microstructure was consistently fine-grained and constant in hardness throughout the specimen's height. The basal texture was also more noticeable in the fusion zone. In a finding, it was discovered that forged AZ31 alloy and AZ31 alloy produced by WAAM had equivalent yield strengths and ultimate tensile strengths [54].

7.9 APPLICATIONS OF WAAM

The creation of large structural parts is being revolutionised by WAAM technology. It is becoming an increasingly popular option for marine propellers, aerospace components, and automotive engineering applications because it provides advantages over other metal AM techniques for similar applications with unmatched efficiency and cost advantages. In order for WAAM to function, metal wire must be quickly deposited onto a substrate in accordance with a 3D design. The deposited layers are

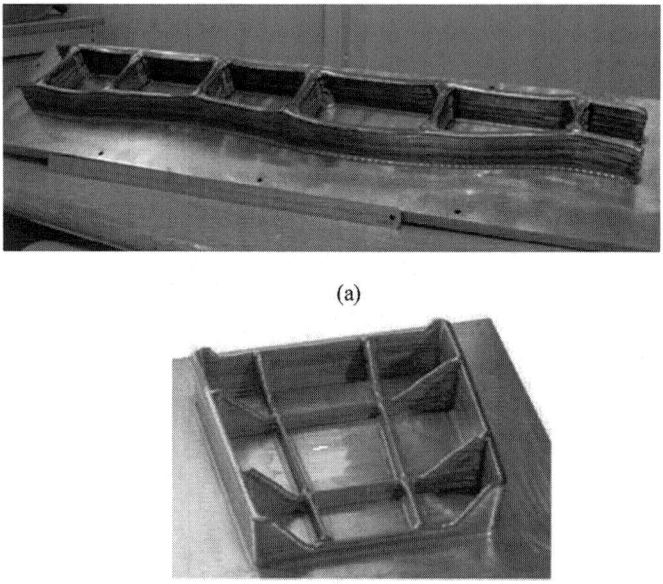

(a)

(b)

FIGURE 7.8 Titanium WAAM products produced by Cranfield University [55]. (a) The wing spar and (b) The landing gear.

then fused together. This layer-by-layer method can produce large parts without the use of additional tooling and is up to ten times or more faster than most other AM techniques. Due to the low cost of the materials used and the lack of post-processing requirements, WAAM is also cost-effective. Additionally, WAAM provides a high level of repeatability and accuracy, making it perfect for the production of intricate and detailed components. With these benefits, WAAM technology is positioned to take centre stage as a crucial tool for the manufacturing of large structural components, enabling increased productivity and cost savings. The European Space Agency recognises WAAM as a low-energy, environmentally friendly production method. In recent years, many research institutions have successfully produced industrial goods using their WAAM systems and have significantly improved them. As shown in Figure 7.8 [55], a Ti-6Al-4V-made 1.2 m wing spar, an external landing gear, and several other frames are examples of successfully manufactured parts. They were able to assemble titanium alloy parts with mechanical characteristics that were exact duplicates of those of common materials. In addition, as shown in Figure 7.9, a 6 m-long aluminium alloy spar has been constructed utilising the WAAM process. When compared to conventional machining, this method can save up to 70% on raw materials [56].

FIGURE 7.9 Al alloy spar having a length of 6 m produced by Cranfield University [56].

7.10 CONCLUSION

The recent technological developments in the WAAM process have been thoroughly reviewed in this chapter. WAAM in non-ferrous materials, specifically alloys of aluminium, nickel, titanium, and magnesium, is discussed along with the basic concepts of AM and then WAAM technologies. The demand for AM and WAAM of Mg and Al alloys is anticipated to rise with the advent of the fourth industrial revolution and the implementation of environmental policies. There are still many areas that require additional research and development, despite the fact that WAAM of Mg and Al alloys is still a relatively new manufacturing technique in comparison to others. It is crucial to conduct further investigation into the fundamental physical and chemical metallurgical mechanisms associated with the WAAM process to provide recommendations for process optimisation and control. This deeper understanding will enable informed decisions and improvements in the WAAM process. The flaws found in parts manufactured by WAAM directly relate to the material qualities and process parameters. It is essential to create methods or supporting processes to stop the occurrence of defects.

REFERENCES

[1] C. Su, X. Chen, C. Gao, and Y. Wang, "Effect of heat input on microstructure and mechanical properties of Al-Mg alloys fabricated by WAAM," *Appl. Surf. Sci.*, vol. 486, pp. 431–440, 2019, 10.1016/j.apsusc.2019.04.255

[2] "ASTM International. F2792-12a – Standard Terminology for Additive Manufacturing Technologies," *Rapid. Manuf. Assoc.*, pp. 10–12, 2013, 10.1520/F2792-12A.2

[3] O. Ivanova, C. Williams, and T. Campbell, "Additive manufacturing (AM) and nanotechnology: Promises and challenges," *Rapid Prototyp. J.*, vol. 19, pp. 353–364, 2013, 10.1108/RPJ-12-2011-0127

[4] M. Dinovitzer, X. Chen, J. Laliberte, X. Huang, and H. Frei, "Effect of wire and arc additive manufacturing (WAAM) process parameters on bead geometry and microstructure," *Addit. Manuf.*, vol. 26, pp. 138–146, 2019, 10.1016/j.addma.2018.12.013

[5] S. W. Williams, F. Martina, A. C. Addison, J. Ding, G. Pardal, and P. Colegrove, "Wire + arc additive manufacturing," *Mater. Sci. Technol. (United Kingdom)*, vol. 32, pp. 641–647, 2016, 10.1179/1743284715Y.0000000073

[6] T. Deb Roy, H. L. Wei, J. S. Zuback, T. Mukherjee, J. W. Elmer, J. O. Milewski et al., "Additive manufacturing of metallic components—Process, structure and properties," *Prog. Mater. Sci.*, vol. 92, pp. 112–224, 2018, 10.1016/j.pmatsci.2017.10.001

[7] S. Reddy, V. Madhava, and C. S. Reddy, "3-D printing technologies and processes—A review international organization of scientific research," *IOSR J. Eng.*, vol. 7, pp. 1–14, 2017.

[8] J. M. Flynn, A. Shokrani, S. T. Newman, and V. Dhokia, "Hybrid additive and subtractive machine tools—Research and industrial developments," *Int. J. Mach. Tools Manuf.*, vol. 101, pp. 79–101, 2016, 10.1016/j.ijmachtools.2015.11.007

[9] P. K. Gokuldoss, S. Kolla, and J. Eckert, "Additive manufacturing processes: Selective laser melting, electron beam melting and binder jetting-selection guidelines," *Materials (Basel)*, 2017, 10.3390/ma10060672

[10] H. Bikas, P. Stavropoulos, and G. Chryssolouris, "Additive manufacturing methods and modeling approaches: A critical review," *Int. J. Adv. Manuf. Technol.*, vol. 83, pp. 389–405, 2016, 10.1007/s00170-015-7576-2

[11] S. Ford, and M. Despeisse, "Additive manufacturing and sustainability: An exploratory study of the advantages and challenges," *J. Clean. Prod.*, vol. 137, pp. 1573–1587, 2016, 10.1016/j.jclepro.2016.04.150

[12] W. Gao et al., "The status, challenges, and future of additive manufacturing in engineering," *CAD Comput. Aided Des.*, vol. 69, pp. 65–89, 2015, 10.1016/j.cad.2015.04.001

[13] J. D. Spencer, P. M. Dickens, and C. M. Wykes, "Rapid prototyping of metal parts by three-dimensional welding," *Proc. Inst. Mech. Eng. Part B J. Eng. Manuf.*, vol. 212, no. 3, pp. 175–182, 1998, 10.1243/0954405981515590

[14] Y. M. Zhang, Y. Chen, P. Li, and A. T. Male, "Weld deposition-based rapid prototyping: A preliminary study," *J. Mater. Process. Technol.*, vol. 135, no. 2–3, pp. 347–357, 2003, 10.1016/S0924-0136(02)00867-1

[15] J. H. Ouyang, H. Wang, and R. Kovacevic, "Rapid prototyping of 5356-aluminum alloy based on variable polarity gas tungsten arc welding: Process control and microstructure," *Mater. Manuf. Process.*, vol. 17, no. 1, pp. 103–124, 2002, 10.1081/AMP-120002801

[16] P. Kazanas, P. Deherkar, P. Almeida, H. Lockett, and S. Williams, "Fabrication of geometrical features using wire and arc additive manufacture," *Proc. Inst. Mech. Eng. Part B J. Eng. Manuf.*, vol. 226, no. 6, pp. 1042–1051, 2012, 10.1177/0954405412437126

[17] C. R. Cunningham, J. M. Flynn, A. Shokrani, V. Dhokia, and S. T. Newman, "Invited review article: Strategies and processes for high quality wire arc additive manufacturing," *Addit. Manuf.*, vol. 22, no. June, pp. 672–686, 2018, 10.1016/j.addma.2018.06.020

[18] A. Horgar, H. Fostervoll, B. Nyhus, X. Ren, M. Eriksson, and O. M. Akselsen, "Additive manufacturing using WAAM with AA5183 wire," *J. Mater. Process. Technol.*, vol. 259, no. April, pp. 68–74, 2018, 10.1016/j.jmatprotec.2018.04.014

[19] J. L. Prado-Cerqueira, J. L. Diéguez, and A. M. Camacho, "Preliminary development of a wire and arc additive manufacturing system (WAAM)," *Procedia Manuf.*, vol. 13, pp. 895–902, 2017, 10.1016/j.promfg.2017.09.154

[20] D. Ding, Z. Pan, D. Cuiuri, and H. Li, "Wire-feed additive manufacturing of metal components: Technologies, developments and future interests," *Int. J. Adv. Manuf. Technol.*, vol. 81, no. 1–4, pp. 465–481, 2015, 10.1007/s00170-015-7077-3

[21] W. J. Sames, F. A. List, S. Pannala, R. R. Dehoff, and S. S. Babu, "The metallurgy and processing science of metal additive manufacturing," *Int. Mater. Rev.*, vol. 61, no. 5, pp. 315–360, 2016, 10.1080/09506608.2015.1116649

[22] V. Bhavar, P. Kattire, V. Patil, S. Khot, K. Gujar, and R. Singh, "A review on powder bed fusion technology of metal additive manufacturing," *Addit. Manuf. Handb. Prod. Dev. Def. Ind.*, 2017, 10.1201/9781315119106

[23] B. Dutta, and F. H. Sam Froes, "The additive manufacturing (AM) of titanium alloys," *Met. Powder Rep.*, vol. 72, no. 2, pp. 96–106, 2017, 10.1016/j.mprp.2016.12.062

[24] A. Ivántabernero, P. Á. Paskual, and A. Suárez, "Study on arc welding processes for high deposition rate additive manufacturing," *Procedia CIRP*, vol. 68, no. April, pp. 358–362, 2018, 10.1016/j.procir.2017.12.095

[25] L. P. Raut, and R. V. Taiwade, "Wire arc additive manufacturing: A comprehensive review and research directions," *J. of Materi Eng and Perform*, vol. 30, pp. 4768–4791, 2021, 10.1007/s11665-021-05871-5

[26] X. Xiong, H. Zhang, and G. Wang, "Metal direct prototyping by using hybrid plasma deposition and milling," *J. Mater. Process. Technol.*, vol. 209, no. 1, pp. 124–130, 2009, 10.1016/j.jmatprotec.2008.01.059

[27] J. Ding et al., "Thermo-mechanical analysis of wire and arc additive layer manufacturing process on large multi-layer parts," *Comput. Mater. Sci.*, vol. 50, no. 12, pp. 3315–3322, 2011, 10.1016/j.commatsci.2011.06.023

[28] P. M. Dickens, M. S. Pridham, R. C. Cobb, I. Gibson, and G. Dixon, "Rapid prototyping using 3-D welding," *Solid Free. Fabr. Proc.*, pp. 280–290, 1992, [Online]. Available: https://repositories.lib.utexas.edu/handle/2152/64409

[29] S. Ríos, P. A. Colegrove, and S. W. Williams, "Metal transfer modes in plasma wire + arc additive manufacture," *J. Mater. Process. Technol.*, vol. 264, pp. 45–54, 2019, 10.1016/j.jmatprotec.2018.08.043

[30] J. Xiong, G. Zhang, Z. Qiu, and Y. Li, "Vision-sensing and bead width control of a single-bead multi-layer part: Material and energy savings in GMAW-based rapid manufacturing," *J. Clean. Prod.*, vol. 41, pp. 82–88, 2013, 10.1016/j.jclepro.2012.10.009

[31] J. Levesque, A. Shah, S. Ekhtiari, J. Yan, P. Thornley, and D. Williams, "Three-dimensional printing in orthopaedic surgery: A scoping review," *EFORT Open Rev.*, vol. 5, pp. 430–441, 2020, 10.1302/2058-5241.5.190024

[32] N. Rosli, M. R. Alkahari, F. Ramli, and M. Fadzli, "Influence of process parameter on the height deviation of weld bead in wire arc additive manufacturing," *Int. J. Mech. Prod. Eng. Res. Dev.*, vol. 10, pp. 1165–1176, 2020, 10.24247/ijmperdjun2020101

[33] P. A. Colegrove et al., "Microstructure and residual stress improvement in wire and arc additively manufactured parts through high-pressure rolling," *J. Mater. Process. Technol.*, vol. 213, no. 10, pp. 1782–1791, 2013, 10.1016/j.jmatprotec.2013.04.012

[34] X. Chen, J. Li, X. Cheng, B. He, H. Wang, and Z. Huang, "Microstructure and mechanical properties of the austenitic stainless steel 316L fabricated by gas metal arc additive manufacturing," *Mater. Sci. Eng. A*, vol. 703, pp. 567–577, 2017, 10.1016/j.msea.2017.05.024

[35] F. Tang, Y. Luo, Y. Cai et al. "Arc length identification based on arc acoustic signals in GTA-WAAM process," *Int J Adv Manuf Technol*, vol. 118, pp. 1553–1563, 2022, 10.1007/s00170-021-08044-9

[36] B. Wu et al., "A review of the wire arc additive manufacturing of metals: Properties, defects and quality improvement," *J. Manuf. Process.*, vol. 35, pp. 127–139, 2018, 10.1016/j.jmapro.2018.08.001

[37] C. Zhang, G. Li, M. Gao, J. Yan, and X. Y. Zeng, "Microstructure and process characterization of laser-cold metal transfer hybrid welding of AA6061 aluminum alloy," *Int. J. Adv. Manuf. Technol.*, vol. 68, no. 5–8, pp. 1253–1260, 2013, 10.1007/s00170-013-4916-y

[38] N. Havar, S. Bhondwe, V. A. Mahajan, and R. Dhoot, "Recent advances in post systems: A review," *J. Appl. Den. Med. Sci.*, vol. 1, p. 3, 2016

[39] C. Shen, Z. Pan, Y. Ma, D. Cuiuri, and H. Li, "Fabrication of iron-rich Fe-Al intermetallics using the wire-arc additive manufacturing process," *Addit. Manuf.*, vol. 7, pp. 20–26, 2015, 10.1016/j.addma.2015.06.001

[40] N. T. Aboulkhair, I. Maskery, C. Tuck, I. Ashcroft, and N. M. Everitt, "The microstructure and mechanical properties of selectively laser melted AlSi10Mg: The effect of a conventional T6-like heat treatment," *Mater. Sci. Eng. A*, vol. 667, pp. 139–146, 2016, 10.1016/j.msea.2016.04.092

[41] R. Karunakaran, S. Ortgies, A. Tamayol, F. Bobaru, and M. P. Sealy, "Additive manufacturing of magnesium alloys," *Bioact. Mater.*, vol. 5, no. 1, pp. 44–54, 2020, 10.1016/j.bioactmat.2019.12.004

[42] J. F. Wang, Q. J. Sun, H. Wang, J. P. Liu, and J. C. Feng, "Effect of location on microstructure and mechanical properties of additive layer manufactured Inconel 625 using gas tungsten arc welding," *Mater. Sci. Eng. A*, vol. 676, pp. 395–405, 2016, 10.1016/j.msea.2016.09.015

[43] J. H. Park, M. Cheepu, and S. M. Cho, "Analysis and characterization of the weld pool and bead geometry of inconel 625 super-TIG welds," *Metals (Basel)*, vol. 10, no. 3, pp. 1–11, 2020, 10.3390/met10030365

[44] W. Yangfan, C. Xizhang, and S. Chuanchu, "Microstructure and mechanical properties of Inconel 625 fabricated by wire-arc additive manufacturing," *Surf. Coatings Technol.*, vol. 374, no. May, pp. 116–123, 2019, 10.1016/j.surfcoat.2019.05.079

[45] B. A. Szost et al., "A comparative study of additive manufacturing techniques: Residual stress and microstructural analysis of CLAD and WAAM printed Ti-6Al-4V components," *Mater. Des.*, vol. 89, pp. 559–567, 2016, 10.1016/j.matdes.2015.09.115

[46] C. Qiu, G. A. Ravi, C. Dance, A. Ranson, S. Dilworth, and M. M. Attallah, "Fabrication of large Ti-6Al-4V structures by direct laser deposition," *J. Alloys Compd.*, vol. 629, pp. 351–361, 2015, 10.1016/j.jallcom.2014.12.234

[47] F. Wang, S. Williams, and M. Rush, "Morphology investigation on direct current pulsed gas tungsten arc welded additive layer manufactured Ti6Al4V alloy," *Int. J. Adv. Manuf. Technol.*, vol. 57, no. 5–8, pp. 597–603, 2011, 10.1007/s00170-011-3299-1

[48] B. E. Carroll, T. A. Palmer, and A. M. Beese, "Anisotropic tensile behavior of Ti-6Al-4V components fabricated with directed energy deposition additive manufacturing," *Acta Mater.*, vol. 87, pp. 309–320, 2015, 10.1016/j.actamat.2014.12.054

[49] M. Kobayashi, Y. Dorce, H. Toda, and H. Horikawa, "Effect of local volume fraction of microporosity on tensile properties in Al-Si-Mg cast alloy," *Mater. Sci. Technol.*, vol. 26, no. 8, pp. 962–967, 2010, 10.1179/174328409X441283

[50] F. Chai, D. Zhang, and Y. Li, "Effect of thermal history on microstructures and mechanical properties of AZ31 magnesium alloy prepared by friction stir processing," *Materials (Basel)*, vol. 7, no. 3, pp. 1573–1589, 2014, 10.3390/ma7031573

[51] Y. Guo, H. Pan, L. Ren, and G. Quan, "Microstructure and mechanical properties of wire arc additively manufactured AZ80M magnesium alloy," *Mater. Lett.*, vol. 247, pp. 4–6, 2019, 10.1016/j.matlet.2019.03.063

[52] D. A. Martinez Holguin, S. Han, and N. P. Kim, "Magnesium alloy 3D printing by wire and arc additive manufacturing (WAAM)," *MRS Adv.*, vol. 3, no. 49, pp. 2959–2964, 2018, 10.1557/adv.2018.553

[53] T. Klein, A. Arnoldt, M. Schnall, and S. Gneiger, Microstructure formation and mechanical properties of a wire-arc additive manufactured magnesium alloy, *JOM*, vol. 73, no. 4, pp. 1126–1134, 2021, 10.1007/s11837-021-04567-4

[54] "Additive and Subtractive Manufacturing," *Addit. Subtractive Manuf.*, 2019, 10.1515/9783110549775
[55] S. W. Williams, F. Martina, A. C. Addison, J. Ding, G. Pardal, and P. Colegrove, "Wire + arc additive manufacturing," *Mater. Sci. Technol.*, vol. 32, pp. 641–647, 2016, 10.1179/1743284715Y.0000000073
[56] "Have we 3D printed the biggest metal part ever," https://waammat.com/blog/have-we-3d-printed-the-biggest-metal-part-ever. (Accessed 1 December 2021).

8 Functionally Graded Materials via Metal Additive Manufacturing

Juneed Yawar and Muhammad Mursaleen
National Institute of Technology Srinagar, Kashmir, India

8.1 INTRODUCTION

Functionally graded materials (FGMs) are mixtures with locally tailored features and are identifiable by subtle variations in structures and/or microstructures in at least one direction. They are used for various engineering applications like nuclear power, aircraft, petrochemical, and photoelectric. FGM's have been found to impart strength, tenacity, resistance to oxidation, corrosion, and wear, as well as being affordable and lightweight [1–3]. FGM has endlessly varying structures, microstructures, and properties that are effective on environmental needs. In nature, maximum FGM exists via organic and inorganic mediums. Direct energy accumulation skilfully utilises the synthetic medium for the finest modification of the domain. Bones are bosker example, and that is not just FGM but graded direct energy accumulation coming out of the bone molecules, the Hydroxyapatite gem, bone tissue, the Haversian system, and dense bones to fruitfully obey numerous functions like mechanical, physiological, and biological, functions [4–6]. Though Rocks are FGM, and that is specific, the rocks that are below the earth's surface have distinct structures of chemical, spongy, compactness, and eolotropic, expected that they have distinct levels of temperature and pressure [7]. The aim is to spread FGM and to link up the trails facing alike colossal temperature gradients, chemical decaying, fatigue, and so on. FGM finds indispensable uses in aerospace, nuclear power, anti-corrosion, and biomedical implantation [8]. However, advancements have been made in the past few decades, particularly with regard to additive manufacturing. Today's innovative voxel-based 3D printers can distribute and precisely arrange extremely tiny amounts (measured in pico-litres) of liquid-shaped material on a substrate before it transforms into a solid, carrying out the progressive change of phase medium. The inquiry is still ongoing for the design and its characterisation [9–11]. The designs of FGM contain many geometrically created techniques. It is necessary to have a mixture of mediums with a specific plan that is regulated by two or more different mediums. When placed in a web stock state, an orthodox mixture medium exhibits rapid change between separate segments, leading to tension buildup and slicing. Functionally graded direct energy accumulation media, the most recent ideal mixed media, is the ideal medium to meet the needs. FGM has consistent variations in its chemical and microstructural

 DOI: 10.1201/9781003363415-8

properties from one side to the other, and this leads to consistent changes in the connectivity of the medium. FGM is produced using standard forging or casting procedures [12]. FGM can now be produced effectively thanks to additive manufacturing. Self-governing manufacturing process, additive manufacturing (AM) approves precise run of structure and numerous other operations forms at the Nano-sized volume through the layered, targeted accumulation of the material [13–15].

Currently, FGM is produced using additive manufacturing techniques using laser beam AM and wire arc AM. The LAM of FGM provides high-end precision at a high cost with low- to medium-level implementation. Additionally, the printed sample size is constrained by the low heat input of the laser heat system. In contrast, wire arc AM technology has a distinct lead in the preparation of FGM, including low medium and tool costs and sky-scraping accumulation rates, and has a capacity for constructing enormous designs with complete density. Furthermore, wire arc AM technology provides precise structure command by using a feeding system of twin-wire with a predetermined speed ratio of wire feed. Creating a geometrical representation of the microstructure in the FGM section, converting it to the standard tessellation language (STL) file format (breaking the geometry into a series of triangles or polygons), planning the printing path, and fabricating the FGM are all parts of geometrically based FGM design and manufacturing. There are only a few significant flaws in this workflow. While the current geometric mannerly algorithms have difficulty achieving high volume fraction of inclusions, locally high volume fraction of inclusions is bound to occur in FGM's, and the generation of a geometric manner for FGM microstructure is typically time-consuming; furthermore, the overhead of numbers increases noticeably when moving from a geometric approach to a boundary representation (B-Rep) [16]. Finding FGM's effective qualities is the main goal of characterisation, and doing so depends on the characteristics of the phase medium and the gradation functions. FGM's are naturally heterogeneous and anisotropic in comparison to regular mixtures.

Gas and crude oil are transported from the fountainhead to the littoral using cylindrical tubes called marine risers. Because of the serious environmental consequences of their failure, they were constructed with structural and functional requisites in mind. Risers are frequently used in top-tensioned, free-standing, hybrid geometric designs and catenary. Because of the presence of chlorides, carbon dioxide, and hydrogen sulfide, gases in crude oil and gas, a marine riser, which is often attached to a semi-submersible, is vulnerable to severe corrosion [5–8]. The deep-water pipage network is also constantly exposed to a variety of high eco-friendly pressures brought on by wind, waves, currents, ice, and impact. This chapter focuses on various direct energy accumulation methods that could be used to create FGM, wire arc additive manufacturing (WAAM), laser-engineered net shaping/laser metal accumulation (LENS/LMD), and electron beam wire feed AM.

Although mixture media are used as alternatives, their use is seriously threatened by delamination when subjected to both thermal and mechanical loads. For the bonding of metal and ceramic, a graded direct energy accumulation Interlayer demonstrates superiority in both thermal resistance and hardness. However, thermal stresses at the interface under high temperatures and pressure lead to crack formation and cause differences in their coefficients of thermal expansion. FGM avoids that specific

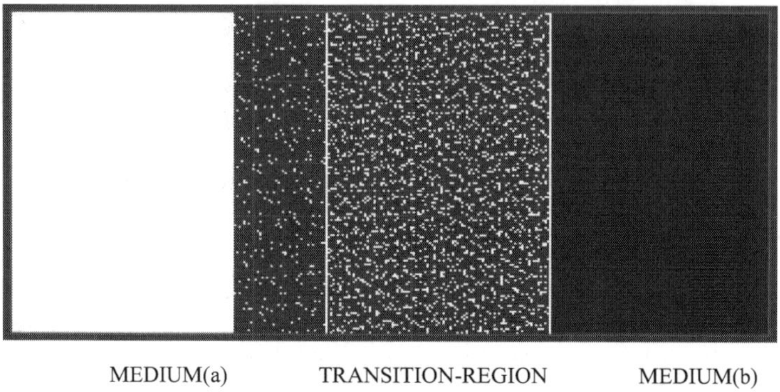

(a)

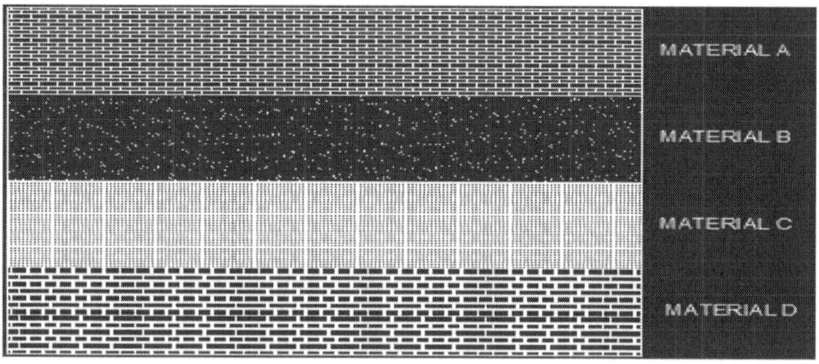

(b)

FIGURE 8.1 Functional graded materials for direct energy accumulation: (a) continuous variation and (b) stepped variation.

issue since the same lacks a distinctive user interface. In a recent study, nickel aluminide and titanium carbide were recommended FGM for pelagic risers. FGM's are classified as either continuous or stepwise in terms of direct energy accumulation according to their uses, as shown in Figure 8.1. In both situations, the volume is altered in a number of qualities, such as density, structure, and microstructure, to produce a property gradient. Titanium, nickel-based, and stainless steel super-alloys are used flourishingly to create metallic FGM' utilising AM processes. FGM was initially created using traditional methods such as centrifugal casting, spark plasma sintering, laser accumulation, and powder metallurgy. However, a few AM methods, such as electron beam melting, selective laser melting (SLM), wire arc AM, and direct laser melting process, have recently made significant strides.

8.2 CLASSIFICATION OF FGM

FGM is classified based on gradient type; the classification is based upon various chemical compositions used in the material depending upon the end usage and application of the material, like high hardness, fatigue strength, and high endurance limit.

FGM is classified based on microstructure gradient type; the classification is done based on the arrangement of atoms, ions, and molecules.

FGM is classified based on porosity type; the classification is done based on porosity ratio and void spaces needed for electrical and heat transfer applications.

FGM is classified based on ceramic metal type; the classification is done based on the ceramic material incorporated needed for specific applications like high temperature and high hardness.

FGM is classified based on metal-metal type; the classification is done based on the metal-metal interaction needed for specific applications.

FGM is classified based on metal-polymer type; the classification is done based on the metal-polymer interaction needed for biomedical applications. Furthermore, the classification is based on the gradient type and constituent type, as shown in Figure 8.2.

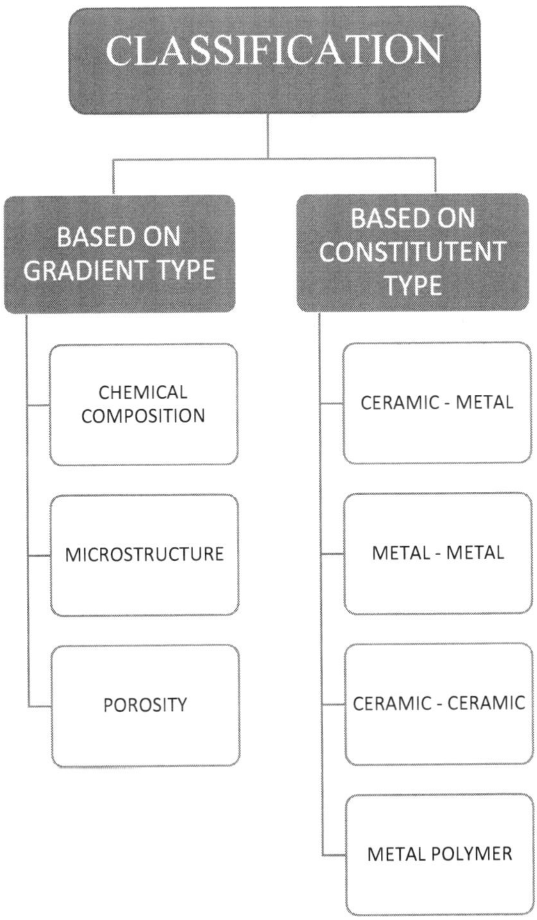

FIGURE 8.2 Classification of FGM material.

8.3 TECHNIQUES OF FABRICATION FOR FGM

The solid-state process means manufacturing materials using solid-state procedures for AM.

The liquid state process incorporates the melted materials used for AM.

The vapour deposition process involves vapourising the material at elevated temperatures and subsequently depositing the same. Figure 8.3 shows the detailed AM processes.

8.4 APPLICATIONS OF FUNCTIONALLY GRADIENT MATERIALS

Materials play a crucial role in industrial production; with the improvement of social productivity, the requirement for materials is also improved. The following flow chart further explains the process. Figure 8.4 shows the detailed applications of FGM materials in various fields.

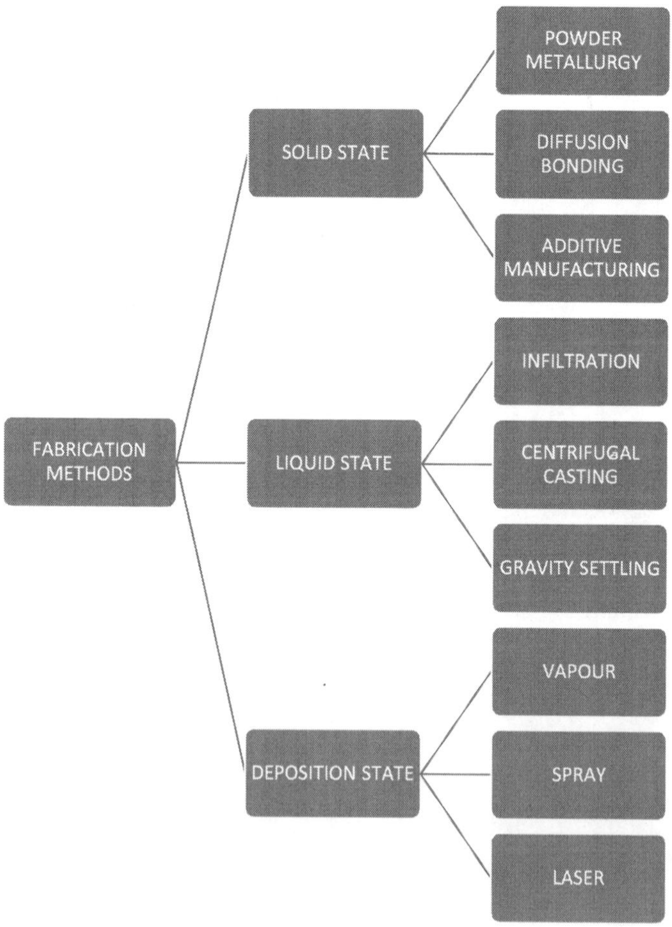

FIGURE 8.3 Fabrication methods used in AM.

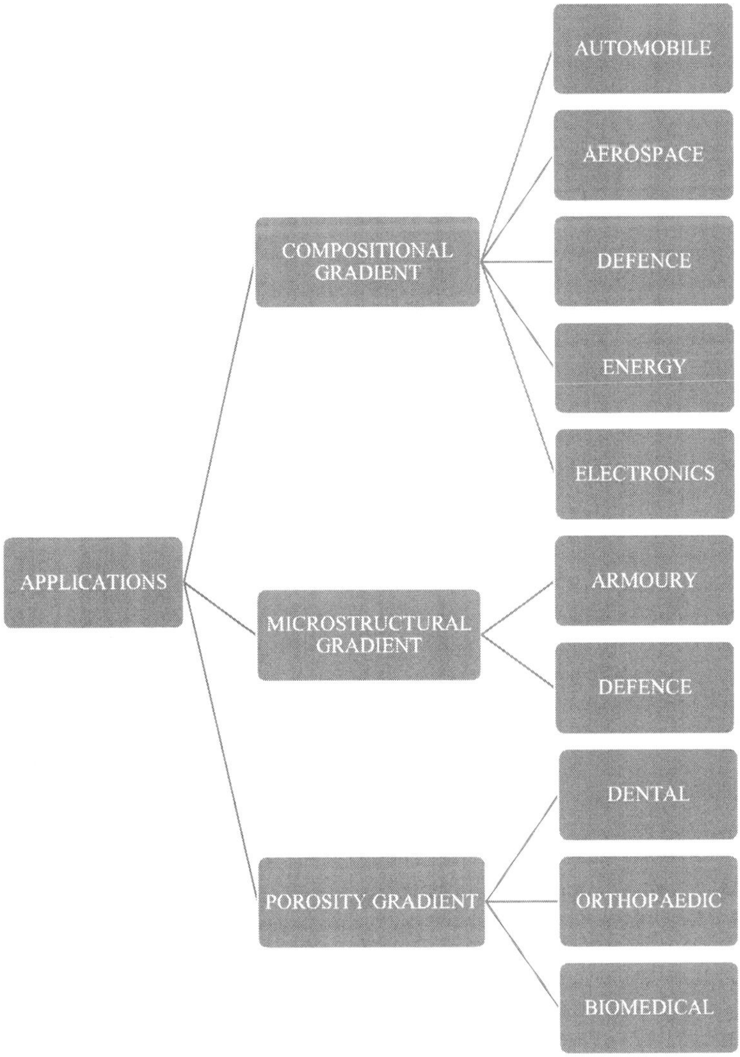

FIGURE 8.4 Application of FGM.

The compositional gradient is used for automobile, aerospace, and chassis body manufacturing, which has been found to produce hard tensile rigid structures.

In the defence industry, they are used to produce arms. They are also used to produce body armour and bullet shields. FGM is used in prosthodontic teeth systems and in bone-cartilage system repair and replacements.

8.5 USING AM TECHNIQUES TO CREATE FGM MADE OF METAL

AM was primarily known as fast prototyping or 3D printing and was used for a very long term to create elaborate blueprints of building up feedstocks layer by layer. AM is an umbrella term for a variety of non-material removal processes, including wire arc AM, electron beam melting, selective laser sintering (SLS), SLM, and LENS. Materials of a broad range of (ceramics, rubber, plastics, metals, and so on) are used for AM processes. AM technology is first too expensive and unworkable for many industries' uses. However, as the twenty-first century came to a close, costs drastically decreased, allowing AM to focus on a number of zones. When juxtaposed with conventional producing techniques, AM has some distinct advantages, like lessening environmental bump, reducing waste, ameliorating material properties, and ecological and shorter lead times. Recent improvements in the process have made it possible to manufacture parts that are as strong as parts made the traditional way. There are several different AM processes, depending on the feedstock and heat system heat source. To meet the fabrication challenges of the web medium, solid-state AM techniques (such as hybrid additive manufacturing (HAM), melt-based additive manufacturing (MB-AM), ultrasonic additive manufacturing (UAM), and friction stir additive manufacturing (FSAM) procedures) (like direct energy accumulation and power bed fusion) have also been developed. Recent developments have enabled the integration of multiple media into a single print component. Component failures are caused by the challenges posed by the medium, techniques, and process parameters. In order to produce micro-proto epitomes to nearly net-shaped components, the fabrication of multi-material graded direct energy accumulation segments necessitates careful consideration of the appropriate material combinations, process features, and post-processing technologies. The investigation of AM's occurrence of FGM has just started and is still in its early stages. Among all the AM processes, certain direct energy accumulation and powder bed fusion (PBF) techniques are effectively used to fabricate metal FGM. Therefore, the importance of direct energy. This chapter examines the accumulation and PBF methods used to create metal FGM components.

8.5.1 AM Method Based on Melting

The heating system that melts the filling is a broad way to categorise AM processes. While solid-state AM adds the material through frictional heat, velocity, and pressure, melt-based AM uses sources of heat, for example, electric arc, laser, and electron beam. This section discusses the PBF technique used for FGM as well as the potential direct energy accumulation.

8.5.1.1 PBF

PBD is an in-depth researched AM technique that layer-by-layer performs Selective sintering or melting of a (fine piled powdered media direct energy accumulation one earlier layer consecutively) powder bed to fabricate a 3D unit. PBF techniques can produce additional working web shapes from various varieties of materials and need little to no support structures. PBF methods also provide other benefits, such as a superior buy a fly ratio that makes it possible to produce FGM's and recycle un-melted powders. The PBF process has a number of significant drawbacks when producing FGM, including the need for a large amount of powder material to direct energy accumulation to build secure and nearby of the useless segment up to the final stage, difficulty in managing the cooling and heating rate, and a cap on the size of wedge. PBF and direct energy accumulation are both AM melt-based, so similar flaws like thermal stress development, lack of fusion flaws, element vapourisation, pore formation, and crack formation can occur. In powder bed material, rollout occurs ahead of melting, but in the case of direct energy accumulation, melting and accumulation occur concurrently; the accumulation efficiency of PBF methods is much lower than that of direct energy accumulation techniques.

8.5.1.2 Electron Beam Melting (EBM)

EBM is dependent on the electrical conductivity of the feeding medium, a conceptual representation of the EBM procedure, as shown in Figure 8.5. When the powder bed is scanned in an EBM process, a higher energy electron beam serves as a heat source, melting the powder to create a build. The process is repeated layer by layer to manufacture completely dense components. Steel and aluminum segments with a thickness ranging from 0.01 mm to 250 mm can be joined using electron beam manufacturing having vacuum conditions. The problem associated with the EBM is the medium powder switchover during fabrication for multi-material production. When compared to the SLM process, the EBM process produces the murkiest spongy composition accompanied by no defects and notably less leftover stress. Although based

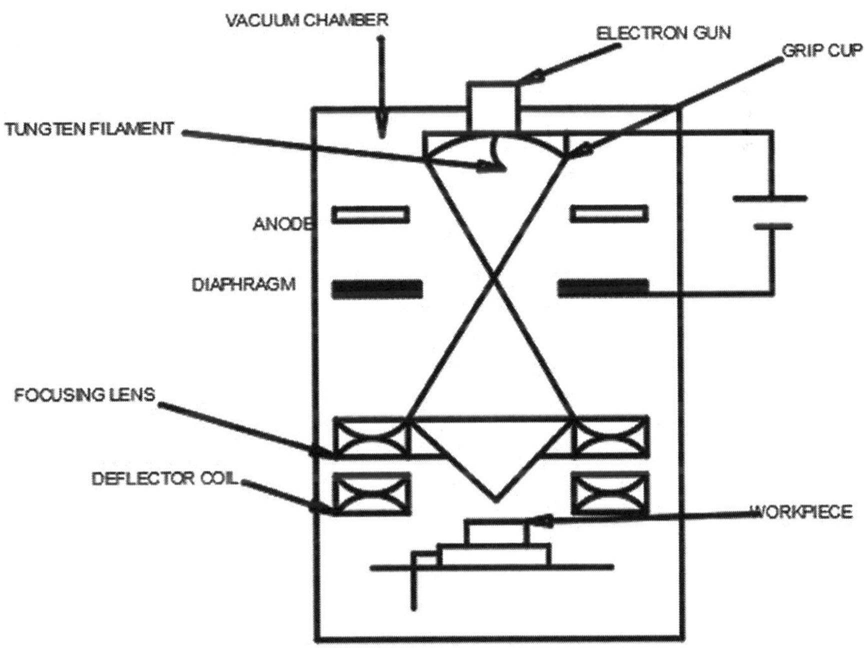

FIGURE 8.5 EBM.

on the laser melting procedure, it uses small spots of laser and fine covering to produce a preferable top glaze than the EBM, heating followed by the softening stage diminishes the thermal gradient and keeps away the development of heat fracture. Even the EBM method has several benefits, including less scrap powder recovery without changing the chemical or physical characteristics of the material, making it universally adaptable for the production of aerospace components. The electron beam is bent by the magnetic field that has been retained in the magnetic medium. The EBM process requires a thorough cleaning of components. The strength of the segments made using EBM is on par with that of conventionally created counter segments.

8.5.1.3 Direct Energy Accumulation (DED)

An essential AM procedure known as direct energy accumulation is frequently used to construct or refurbish top-quality and top-value items. Feedstock for the direct energy accumulation process can be either metal wire or metal powder. A small-vigorous heat system (such as a laser beam, electric arc, and electron beam) is added to soften the feedstock medium after metal wire or powder has been introduced. Layer by layer, the liquid material is spread across the substrate. Manufacturing the FGM components for the direct energy accumulation method involves alloying and multiple-medium feeding procedures. Where the geometric twist is the objectionable component, the direct energy accumulation procedure is largely applicable. Another noteworthy benefit of the AM process is its higher removal rate, a wide range of feedstock mediums, maximal medium application (wire-based), and improved mechanical properties compared to power bed fusion techniques. For the manufacture of FGM, the direct energy accumulation procedures are selected above other AM for the reason of shorter release times, effective material utilisation, and increased flexibility. Defects like rough surfaces, reduced dimensional accuracy, and brittle intermetallic compound layer formations (primarily caused by alloy animosity and insolubility) can develop during the direct energy accumulation processing of FGM due to process variables like scanning speed, layer thickness, energy density, etc. Its Investigated high thermal stresses, waviness, and rough surfaces of the direct energy accumulation treated components typically necessitate post-processing.

8.5.1.4 Laser-Engineered Net Shaping

The terms "laser-engineered net shaping" (LENS), laser metal accumulation (LMD), "laser cladding," and "LENS" are trademarks. The fundamental technology frequently used in AM is LENS/LMD. LMD/LENS cutting-edge AM method was created to create near-net-shaped components that only need a few finishing steps, including final machining. A LENS schematic diagram. The optical setups, laser source, build chamber, build bed, and powder feeder are the main elements of the LENS system. The heat source for LENS/LMD is a laser beam, while the feedstock medium is a powder. A high-energy-density laser beam is used to focus on the substrate plate to generate the molten pool, and feedstock medium is constantly injected towards the molten pool via the powder delivery unit. Nozzles are coaxial dual powder feeders positioned as LMD/LENS to feed metal powder to the laser beam's focal point. When fabricating FGM using the LENS process, the two materials are typically powdered to a size of about 45 to 150 m and fed separately into a coaxial powder feeder. As the powder is fed from the coaxial feeder to the nozzles, the two materials are then combined in place. The carrier gas is argon. The coaxial feeder spinning speed regulates the powder feed rate. The lens is a particular technique used as an integrated laser light source and powder processing to produce very precise segments. Although LENS and LMD processes are similar, there are a few subtle differences between them. The LENS method employs powder as its feedstock, while the LMD process uses filler wires. Laser metal accumulation has a substantially higher rate of material deposition than LENS. LMD is, hence, better suited for creating broad, thick segments. The distance to the hatch, laser power, powder flow rate, hatch angle, scanning speed, laser absorptivity of the material, thermal properties like melting point, metallurgical traits like crystal structure, chemical affinity, atomic size, etc. Some of the examples of the factors that might affect the performance of a laser are important factors that influence the fabricating quality of FGM through LENS/LMD. Due to advantages including flexible, accurate heating, a tiny area influenced by the heat, the potential to manufacture, fine microstructure, and high relative density segments due to increased internal feature creation, LENS/LMD are frequently used to combine different metals. In order to prevent oxidation brought on by the environment, the LENS/LMD process is normally carried out in a process chamber or chamber filled with inert gas.

8.5.2 Processes for Solid-State Additive Manufacturing

The parent metal is melted using a separate heating source in the beam/arc-based AM. Remaining tensions, elemental losses, elemental segregation, cracking, and sponginess can all arise from the procedure. An extremely inhomogeneous microstructure and anisotropic mechanical behaviour are also the results of the changing thermal cycles. The solid-state AM method is free from heat-induced flaws since it lacks the ability to melt a material.

8.5.2.1 Manufacturing by Friction Stir Additive Manufacturing (FSAM)

A more advanced method of manufacturing than friction stir welding, FSAM creates the junction as a rigid form. The ratio of the peak material temperature to the melting temperature (Tm), which corresponds to 0.6 to 0.9 times the Tm, is what determines the maximum similar temperature on the Kelvin scale. Segments do not face the problems associated with quick solidification, such as hot element segregation, hot cracking, porosity, and increased residual stresses, because melting is not involved in the FSAM process. Dynamic recrystallisation can be used for the FSAM process when materials experience plastic deformation under conditions of high strain rates and high temperatures, as shown in Figure 8.6.

The temperature history of the FSAM process has a sharp peak, much to the thermal history of friction stir welding (FSW). In FSAM, a refined, balanced structure with isotropic mechanical properties and high-angle grain boundary characteristics is used for dynamic recrystallisation. In contrast to the feedstock material, the deposited material in FSAM has smaller grains. Epitaxial solidification in beam-based AM procedures results in highly directed columnar structures with anisotropic

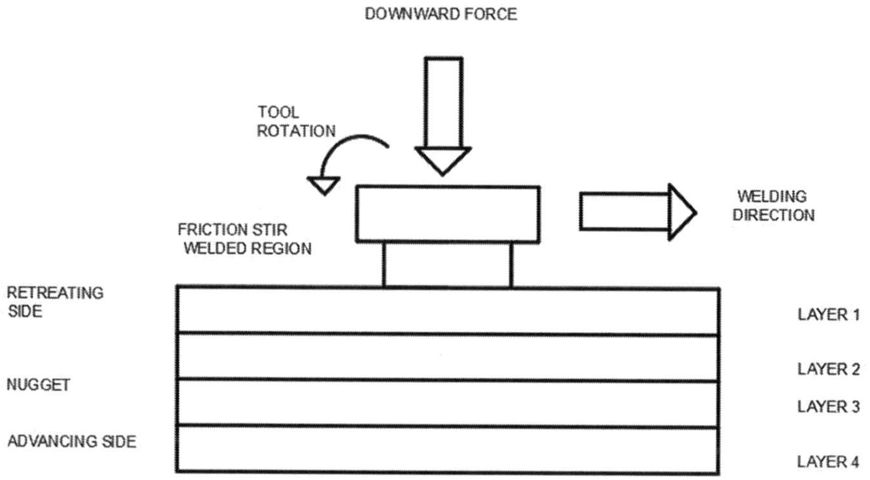

FIGURE 8.6 FSAM.

mechanical characteristics. FSAM is gaining enormous popularity for producing lightweight components. FSAM uses two strategies since it includes layer-by-layer construction.

i. Metallic sheets are layered on top of one another, a revolving FSW tool is moved over their top surfaces, creating frictional heat to link the sheets, and then layers are constructed by laying additional sheets, one at a time, on top of the formerly made layers.
ii. In the method, a powder-filled material is fed into the stirring zone using a hollow FSW tool. This allows the powder particles to soften and direct energy accumulation with previously created layers. The FGM can be produced without having to worry more about the mismatch in coefficient of thermal expansion, residual stress generation, or crack sensitivity because the FSAM process does not include melting. However, only a small amount of instigation work was done in the FSAM of FGM.

8.5.2.2 Ultrasonic Additive Manufacturing (UAM)

The UAM technique was created by Dawn White in 1999 and uses ultrasonic welding to create three-dimensional (3D) parts from a series of ordered foils. The most popular moniker for the FSAM process is ultrasonic consolidation. Over thin metal layers, 20 kHz ultrasonic vibrations are used. The sonotrode's downforce, linear movement, and power supply are all under the direction of the control system. High-frequency vibration-producing transducers and sonotrode horns are the main elements of UAM. The ultrasonic AM technique disrupts the oxide layer of the parent media rather than melts it in order to form a junction. The advantages of FSAM also

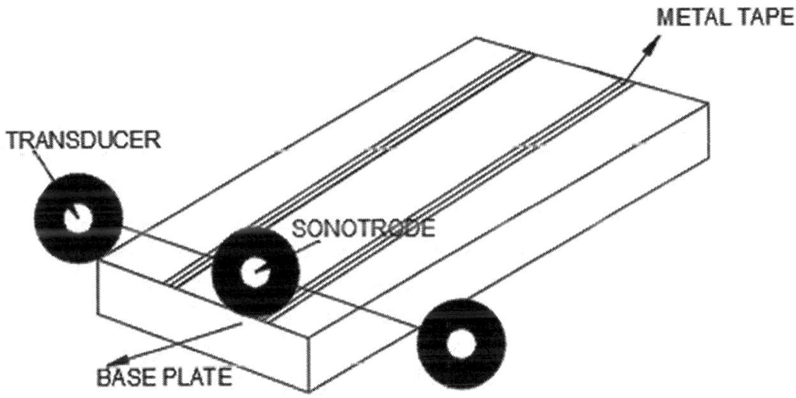

FIGURE 8.7 UAM.

apply to UAM. The UAM method's lower processing temperatures make it possible to fabricate dissimilar material structures, heat-sensitive electronics accessories, dissimilar material structures, joining of aluminum and copper alloys, and smooth heat transport channels, as shown in Figure 8.7.

8.5.2.3 Manufacturing with Cold Spray Additive Manufacturing (CSAM)

Originally a spray coating process, cold spraying has been developed as an AM mode to create separate segments and replace broken ones. High-compression propellant gas (air, nitrogen, helium, etc.) is used in CSAM, forced on a substrate to induce accumulation while also accelerating the feedstock powder at a speed exceeding the critical impact velocity (greater than 300 m/s). Due to CSAM's high kinetic energy dependence rather than thermal energy, the development of residual strains, phase, and oxidation transitions was reduced, as shown in Figure 8.8.

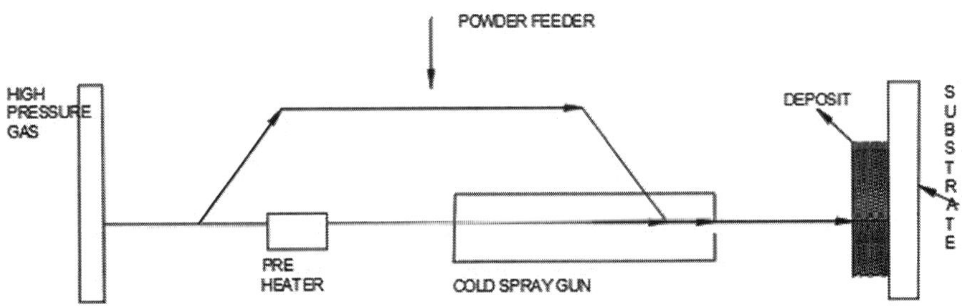

FIGURE 8.8 CSAM.

8.5.3 Hybrid Methods for Additive Manufacturing

In a hybrid additive manufacturing (HAM) process, web segments not produced by a single AM technique are created by combining.

i. Two or more than two AM processes, such as SLM + CSAM, laser + Arc, or EBM + CSAM.
ii. A worming process, such as cold rolling + hot forming, SLM + WAAM, etc.
iii. subtractive constructing processes, such as milling + WAAM, SLM + milling.

8.5.3.1 WAAM, or Wire Arc Additive Manufacturing

Digitally direct energy accumulation WAAM allows the step-by-step insertion of layers of material so as to create 3D components. Switching the material of the wires throughout the accumulation process, WAAM can produce a functional gradient direct energy accumulation material. In the current study, WAAM of 1.2 mm diameter wires of two metals, especially ER 70S-6 carbon manganese steel and ER 2209 duplex stainless steel, is being used to construct necessary FGM. The WAAM process was effectively used to get beyond power bed printer (PBP) restrictions in terms of accumulation rate, equipment, and powder cost during the manufacturing process. As a result, the process of producing FGM is cost-effective because the pricey heat system was removed, and wire feedstock and an electric arc were used in place of the metallic powder, as shown in Figure 8.9. In addition to assisting with a higher accumulation rate, it also promotes quick manufacturing with little material waste. In WAAM, the wire is melted and deposited in layers using an arc-based welding technique [16]. The parameters of the motion guidance and metal transfer are both factors that affect how precisely dimensioned the produced component is. If two distinct metals are used for the wires, switching the wires throughout the accumulation process creates a single FGM segment. When compared to other fabricating methods, WAAM-based FGM fabrication is less expensive in terms of equipment and raw materials.

Vapour accumulation, powder metallurgy, centrifugal casting, and other traditional manufacturing processes are used to create FGM's. For thin-coating FGM's,

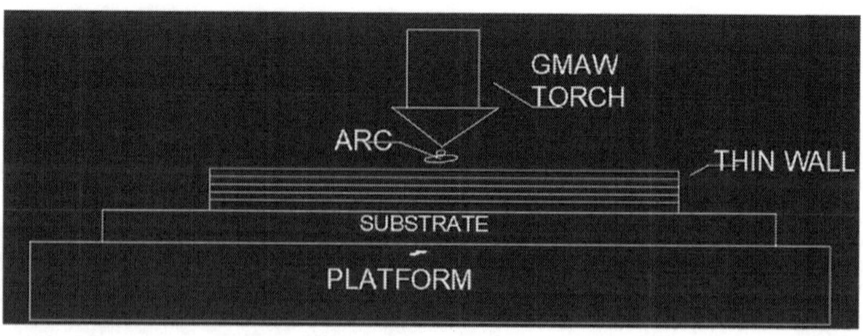

FIGURE 8.9 WAAM.

vapour accumulation is used. Vapour accumulation is not a green method of generating bulk FGM because of its excessive energy use and damaging byproducts. Bulk FGM's are produced via centrifugal casting and powder metallurgy. Powder material manufacturing in accordance with a predetermined dispersion formulation, stacking, and final sintering is required for the creation of FGM's utilising powder metallurgy. Powder metallurgy can only handle stepwise grading since stacking and powder mixing must occur before the final sintering process. Continuous grading is permitted to be accomplished via centrifugal casting. However, only cylindrical FGM sections, like tubes, can be acquired due to hardware restrictions. The lengthy process of the traditional procedures mentioned earlier is caused by the requirement for numerous manufacturing tools and assembly steps. Today, a novel strategy known as AM is piquing enormous interest. In contrast to subtractive manufacturing techniques, AM is defined by the American Society of Testing Materials (ASTM) as the process of joining material to make objects from 3D manner data, typically layer by layer. A 3D computer model of 3D objects is created and divided into thin layers using computer-aided design direct energy accumulation tools. The AM system will then combine all the layers together to produce a 3D part by shaping each layer in accordance with the instructions of the tool path. The ability to create web shapes and customised segments, as well as its shorter lead times, need for less equipment and assembly, and overall lower waste generation, make AM a direct energy accumulation process. Numerous investigation works on metal AM have been conducted as a result of the significant role that metals and their alloys play in our daily lives. For creating laser powder forming (LPF), metallic components and low plasticity burnishing (LPB) AM are the two main exemplars of AM. According to ASTM standards, LPF is one of the seven categories of AM under the heading of direct energy accumulation (direct energy accumulation), and it is also regarded as direct energy accumulation as LMD. In the case of the laser powder-fed procedure, a focused laser beam is utilised to apply a melt pool to a metal substrate. Streams of powder particles are being blown into the melt pool by powder feeders at the same time. Each layer's movement is managed by computer numerical control. Layer by layer, this procedure will be repeated until a 3D metal component is produced. The entire procedure is performed under an inert gas atmosphere to prevent metal oxidation. In LPB, a layer of powder on the powder bed is selectively melted using a laser beam. LPB is often referred to as SLM. Molten powder soon solidifies by joining together. The layer-by-layer laser melting procedure is then repeated after a layer has solidified. A new coating of metal powder is then placed onto the fabrication zone by the roller. As this procedure goes on, a 3D part that is as-built will be produced along with more recyclable, un-melted powder fragments, as shown in Figure 8.10. This explains why waste generation is reduced by subtractive manufacturing processes like AM.

Metal additively made metal FGM's have gained a lot of attention as a result of the rapid expansion of metal AM and metal FGM's, for distinct epitomes of metal FGM's, LMD, and SLM are used. By adjusting the feeding rate for powder distinct epitomes, the LMD technique is effective at managing the chemical structure, but it struggles to regulate other features, such as pores or lattices. In order to produce FGM's that are graded as direct energy accumulation in chemical structure, LMD is a superior option. SLM is a more effective way to accurately control the spatial

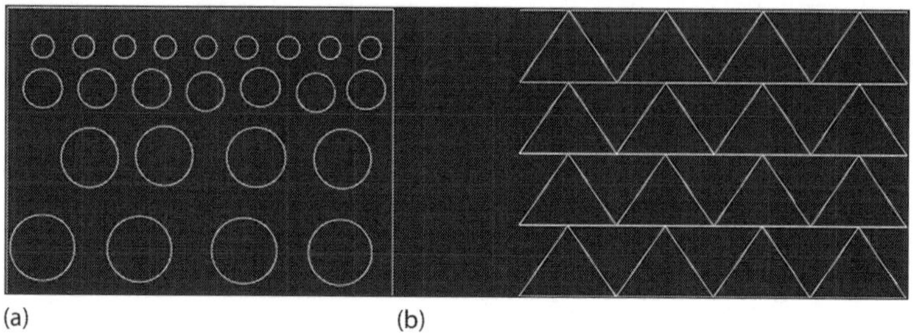

FIGURE 8.10 (a) Structurally graded direct energy accumulation FGM in spongy grading and (b) lattice structure grading.

distribution of material and connect to special structural design requirements for FGM's, such as scaffolds or lattice structures, that are functionally graded for direct energy buildup in spatial geometry. Even if there are now a lot of review publications on FGM's, there is still a paucity of knowledge regarding the AM of metal FGM's. The research presents the investigational work carried out in SLM and LMD producing metal FGM's with grading in structure and structure. The main obstacles to constructing metal FGM are thoroughly debated.

8.5.3.2 Laser Metal Deposition (LMD)

By progressively changing the ratio of distinct metal powders, LMD is able to alter the chemical structure at distinct accumulation places. From layer to layer, it produces a graded direct energy accumulation mixing ratio of metal particles. After fabrication, the graded direct energy accumulation ratio will produce a chemical structure and mechanical properties with a graded direct energy accumulation. This method is very helpful in several production sectors, such as the automobile and aerospace industries, where it is necessary to incorporate multiple exemplars of exceptional mechanical qualities into a single metal component, as shown in Figure 8.11. For instance, a cutting tool needs to have enough root toughness as well as high edge hardness and wear resistance. Sometimes, two good mechanical properties, such as high hardness and high ductility, cannot coexist in a part made of a homogeneous

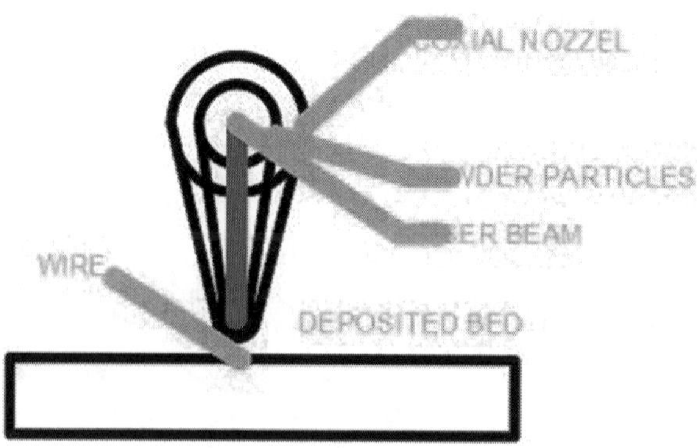

FIGURE 8.11 LMD.

material. LMD can therefore be utilised to produce new designs of metal components by combining two or more media to make metal FGM's that satisfy the particular specifications in discrete parts. LMD will discuss recent important discoveries in the manufacture of metal FGM's in this part. One of the key objectives is to design metal FGM's via LMD with a strengthening phase added gradually. Titanium alloys, especially Ti-6Al-4V, are widely utilised due to their low density, high strength-to-weight ratio, and biocompatibility. A significant amount of research has been done on the practicality of creating Ti-based FGM's reinforced by TiC to boost the hardness and wear resistance of titanium alloys [8]. Earliest successful fabrication of Ti/TiC FGM. With a stepwise gradient of 10% TiC, 20% TiC, and 30% TiC, direct energy accumulation Ti-6Al-4V/TiC was functionally graded by LMD [7]. The relationship between hardness and structural gradient was examined. The Vickers hardness rose from 300 HV1.0 to 600 HV1.0 at the Ti-6Al-4V bottom substrate. The LMD technique was used to create Ti-40vol.%TiC FGM. Distinct laser settings were applied in distinct layers depending on the changes in the material structure. The Ti-40vol.%TiC FGM's hardness increased smoothly as the TiC concentration increased, and no obvious contact could be noticed. FGM with a composition ranging from 100% Ti-6Al-4V to 50% Ti-6Al-4V/50% TiC composed of Ti-6Al-4V/TiC. We compared the wear volumes of samples made using fixed parameters versus samples made using improved parameters. It was discovered that the settings might be adjusted to generate improved wear resistance. Li et al. looked into the relationship between TiC volume and tensile properties. With a 5% TiC volume fraction, the ultimate tensile strength (UTS) of Ti-6Al-4V/TiC FGM increases by about 12.3%. However, the tensile strength of FGM began to decline when the TiC volume fraction started to exceed the direct energy accumulation by 5%.

8.5.3.3 Selective Laser Melting (SLM)

The powder bed–based fabrication method, SLM, is not flexible enough to provide in-flight structural control. Therefore, it is improper to produce FGM's with graded direct energy accumulation chemical structures using SLM. Investigations were carried out in another area of FGM, and direct energy accumulation cellular medium credit goes to the high accuracy of SLM. Direct energy accumulation cellular medium that is functionally graded frequently takes the shape of cells with grading in lattice structures or spongy in three dimensions. SLM is an effective method for producing components with intricate lattice structures that are challenging in standard fabricating. Making moulds for the casting process is excessively time-consuming because of the web's 3D geometry, and assembly is problematic. Solidification and melting of layers may be accurately and selectively controlled by SLM, which allows for the differentiation of solid layers in the end, as shown in Figure 8.12. As opposed to periodic cellular structures, which essentially repeat a unit cell over the 3D space, cellular structures with gradients offer additional desirable properties. They differ from laser melting deposition structurally graded direct energy accumulation FGM's, which typically require fully dense manufacturing to avoid significant flaws in extreme stock states. In some circumstances, the cellular medium will be used in conjunction with hollow or lightweight constructions. The mechanical reaction under quasi-static stock and dynamic impact is a key field for the investigation and development of

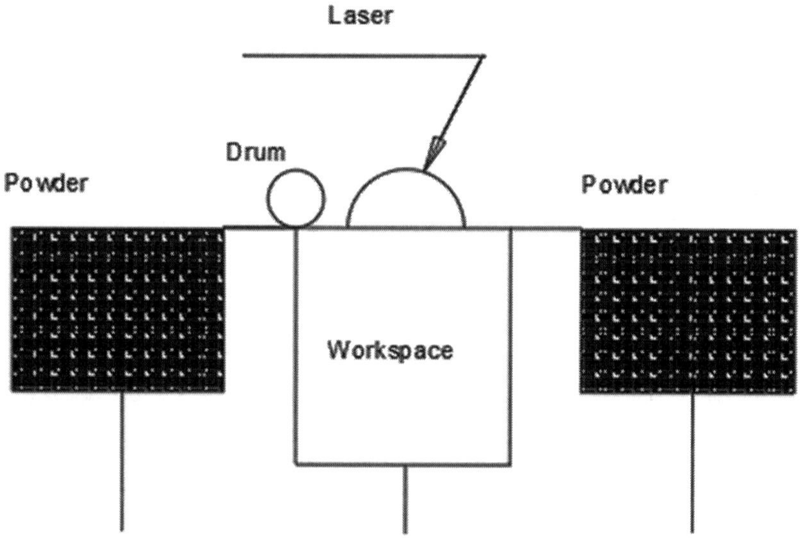

FIGURE 8.12 SLM.

cellular solids. Making lightweight lattice structures for absorbing energy has also gained popularity recently. Lattice structure's distinctive mechanical behaviours and capacity for absorbing energy have been demonstrated by prior investigations. Recently, experimenters have begun to add gradients to these fascinating structures in an effort to uncover additional novel mechanical behaviours. Although there has never been a gradient, there have been numerous designs for cellular structures. We can fabricate lattice structures without the use of moulds because of the inherent high degree of design freedom and quick development of manufacturing precision that SLM can offer. By adjusting the volume fraction in various manufacturing stages, we can also introduce a gradient into those media. We can create direct energy accumulation lattice structures by controlling the 3D geometric design and graded direct energy accumulation spongy segments by changing the building parameters, such as laser power, scanning speed, and hatch spacing.

8.6 CONCLUSION

Particularly, cracking is a frequent problem if the mixture is not blended properly. It has been demonstrated that for particle-reinforced graded direct energy accumulation alloys, a larger particle concentration can lead to wear resistance and greater hardness, but the mark must manage to lessen the number of un-melted particles and to prevent the fracturing roots. Cracks were seen when the TiO_2 concentration varied from 0 to 100%. After that, the maximum TiO_2 concentration was lowered to 90% to create a specimen that wasn't broken. An intermediate transition path between two incompatible materials can occasionally be employed to prevent some intermetallic phases from forming, which are known to easily crack or even shatter. A multi-metallic structure composed of Ti-6Al-4V, V, Cr, Fe, and stainless steel 316 was developed in order to successfully link Ti-6Al-4V and stainless steel 316 without producing a brittle Fe-Ti phase. Powder feeding, a web-based activity, is another key component of LMD that could have an effect on the quality of the final product. The behaviour under an argon gas flow may differ as a result of the variation in particle size and density in this step. The predesigned powder ratio could be broken by powder separation, which would cause a significant departure from the anticipated metal structure. An approach for handling this production problem was examined mathematically and experimentally using the premixed powder flow beneath the carrier gas. Currently, the majority of LMD's FGM segments are test samples. Small test samples must undergo extensive system control development before becoming fully functional components. Therefore, in order to create functional web segments rather than single walls, it is crucial to conduct research on a global approach to LMD path planning. SLM is the most popular metal powder bed AM technique. In addition to SLM, other powder bed–based methods can show their effectiveness in metal FGM's. In order to produce graded direct energy accumulation, another viable method is spongy SLS, particularly for FGM's with graded direct energy accumulation in spongy. Powder particles are only partially melted in SLS, which is the primary distinction between SLS and SLM. More pores will develop between the powder particles as a result of the partial melting. Produced functionally graded direct energy accumulation porous polyamide using SLS. More investigation is still needed into the use of SLS in metal FGM's. Another AM technique based on a powder bed is EBM. The primary distinction between EBM and SLM is that an electron beam, as opposed to a laser beam, is used as the energy source in EBM. Recently, EBM has also been used to create web-graded direct energy accumulation lattice structures with high strength and energy absorption. EBM can be as accurate as SLM.

The main problems with SLM are largely related to design, solid mechanics, and medium extracting. Residual stresses and surface roughness are significant problems for SLM processing. Particularly since pores or micro-scale struts are important ingredients of cellular-graded direct energy accumulation. Surface roughness and residual stress in FGM produced by SLM can have a substantial impact and potentially change the deformation behaviour. When compared to a bulk solid medium, graded direct energy accumulation lattice structures with fine details may be more sensitive to size dimensions. A heat treatment procedure is necessary for SLM segments. When lattice structures with and without heat treatment were evaluated, it was discovered that there were differences in the compressive response between the two. Making the heat treatment process effective is advantageous because if the post-treatment process takes too long, AM will lose some of the advantages of quick manufacturing. The fatigue life of metal segments can be shortened by residual stress as well as surface finish. Understanding how much those elements can influence fatigue behaviour is crucial for graded direct energy accumulation lattice devices used for long-term operation. For graded direct energy accumulation lattice structures up until this point, distinct epitomes of unit cell designs, such as body-centred cubic (BCC), F2 BCC, or rhombic dodecahedron, have existed. The majority of them are founded on experimental methods. Numerical mannerly is developing into a crucial instrument to cut down on medium usage as a result of the quick development of computational techniques. Topology optimisation can be used to find an acceptable unit cell and density distribution design. Designing improved support for structurally graded direct energy accumulation medium with a complex overhang structure or a multidirectional grading is another area where topology optimisation may be used. The primary method for predicting the mechanical response in a distinct stock state is finite element modelling. In contrast to solid bulk segments, cellular structures have a much wider range of potential reactions, and some behaviours are still difficult to simulate. Additionally, SLM precision is thought to contribute to the discrepancy between experimental and numerical results. In other words, the creation of additively made metal FGM's necessitates collaboration across a wide range of disciplines, including numerical modelling, solid mechanics in cellular structures, phase transition, and AM system development.

REFERENCES

1. Sanjeevi Prakash, K., Kannan, A.R. and Shanmugam, N.S., 2023. Additive manufacturing of metal-based functionally graded materials: overview, recent advancements, and challenges. *Journal of the Brazilian Society of Mechanical Sciences and Engineering*, *45*(5), p. 241.
2. Luo, Y., 2023. Voxel-based design and characterization of functionally graded materials. *Results in Materials*, *17*, p. 100375.
3. Xin, D., Yao, X., Zhang, J. and Chen, X., 2023. Fabrication of functionally graded material of 304L stainless steel and Inconel625 by twin-wire plasma arc additive manufacturing. *Journal of Materials Research and Technology*, *23*, pp. 4135–4147.
4. Chandra Sekaran, S., Hari, S. and Amirthalingam, M., 2020. Wire arc additive manufacturing of functionally graded material for marine risers. *Materials Science and Engineering: A*, *792*, p. 139530.
5. Ali, S.I. and Ahmad, S.N., 2023. Tribo-corrosion behavior of Zn-Ni-Cu and Zn-Ni-Cu-TiB2 coated mild steel. *Arabian Journal of Chemistry*, *16*(5), p. 104648.
6. Ali, S.I. and Ahmad, S.N., 2022. Erosion Behavior of Zn-Ni-Cu and Zn-Ni-Cu-TiB2 Coated Mild Steel. *Journal of Materials Engineering and Performance*, *32*, pp. 7406–7417.
7. Ali, S.I. and Ahmad, S.N., 2023. Wear and Corrosion Behavior of Zn-Ni-Cu and Zn-Ni-Cu-TiB2-Coated Mild Steel. *Journal of Materials Engineering and Performance*, *32*, pp. 1–16.
8. Ali, S.I. and Ahmad, S.N., 2022. Corrosion behavior OF HVOF deposited Zn–Ni–Cu and Zn–Ni–Cu–TiB2 coatings on mild steel. *Corrosion Reviews*, *40*(6), pp. 611–624.
9. Srivastava, M. and Rathee, S., 2018. Optimisation of FDM process parameters by Taguchi method for imparting customised properties to components. *Virtual and Physical Prototyping*, *13*(3), pp. 203–210.
10. Rathee, S., Maheshwari, S., Siddiquee, A.N. and Srivastava, M., 2017. Effect of tool plunge depth on reinforcement particles distribution in surface composite fabrication via friction stir processing. *Defence Technology*, *13*(2), pp. 86–91.
11. Rathee, S., Maheshwari, S., Siddiquee, A.N. and Srivastava, M., 2018. A review of recent progress in solid state fabrication of composites and functionally graded systems via friction stir processing. *Critical Reviews in Solid State and Materials Sciences*, *43*(4), pp. 334–366.

12. Srivastava, M., Rathee, S., Maheshwari, S., Noor Siddiquee, A. and Kundra, T.K., 2019. A review on recent progress in solid state friction based metal additive manufacturing: friction stir additive techniques. *Critical Reviews in Solid State and Materials Sciences*, *44*(5), pp. 345–377.
13. Rathee, S., Maheshwari, S. and Siddiquee, A.N., 2018. Issues and strategies in composite fabrication via friction stir processing: A review. *Materials and Manufacturing Processes*, *33*(3), pp. 239–261.
14. Reddy, M.M., Sharma, S., Tharun, D. and Mursaleen, M., 2021. Design and analysis of composite leaf spring under static loading. *AIJR Abstracts*, 110, p. 62.
15. Hajam, S.U.D. and Mursaleen, M., 2021. Design and analysis of an automotive single plate friction clutch using different materials. *AIJR Abstracts*, p. 119.
16. Mushtaq, S., Wani, M.F., Nadeem, M., Najar, K.A. and Mursaleen, M., 2019. A study on friction and wear characteristics of Fe–Cu–Sn alloy with MoS2 as solid lubricant under dry conditions. *Sādhanā*, *44*(12), p. 240.

9 Influence of Manufacturing Parameters on the Improvement of Surface Quality of Wire Arc Additive Manufacturing of Aluminium Alloys

An Overview

Babatunde Olamide Omiyale, Ikeoluwa Ireoluwa Ogedengbe, Temitope Olumide Olugbade, Amos Babatunde Osasona, and Peter Kayode Farayibi

Federal University of Technology, Akure, Nigeria

9.1 INTRODUCTION

Additive manufacturing technology has seen remarkable development over the past few decades [1–3]. Metal additive manufacturing (MAM) has become one of the AM techniques that have witnessed rapid growth in the production of bespoke components for industrial application purposes [1]. Wire arc additive manufacturing (WAAM) technique is fast becoming a core emerging manufacturing technology that has been widely accepted in both academia and industries because of its ability to fabricate large-scale metallic parts for the production of near-net shape components [2] (see Figure 9.1). WAAM technique works by using an electric arc (the heat source) to melt a metal wire, which is subsequently used to fabricate functional metallic components in a layer-by-layer procedure. Considering MAM techniques, WAAM and wire laser additive manufacturing (WLAM) are categories of AM that utilise wire (usually metal) as a material feedstock. The key distinction between these two technologies is the energy source utilised in the wire melting process. While WAAM makes use of an electric arc, WLAM utilises a laser as its energy source. WAAM is a more highly efficient technique of fabricating metallic parts than WLAM. An important benefit of using WAAM is the ability to fabricate

DOI: 10.1201/9781003363415-9

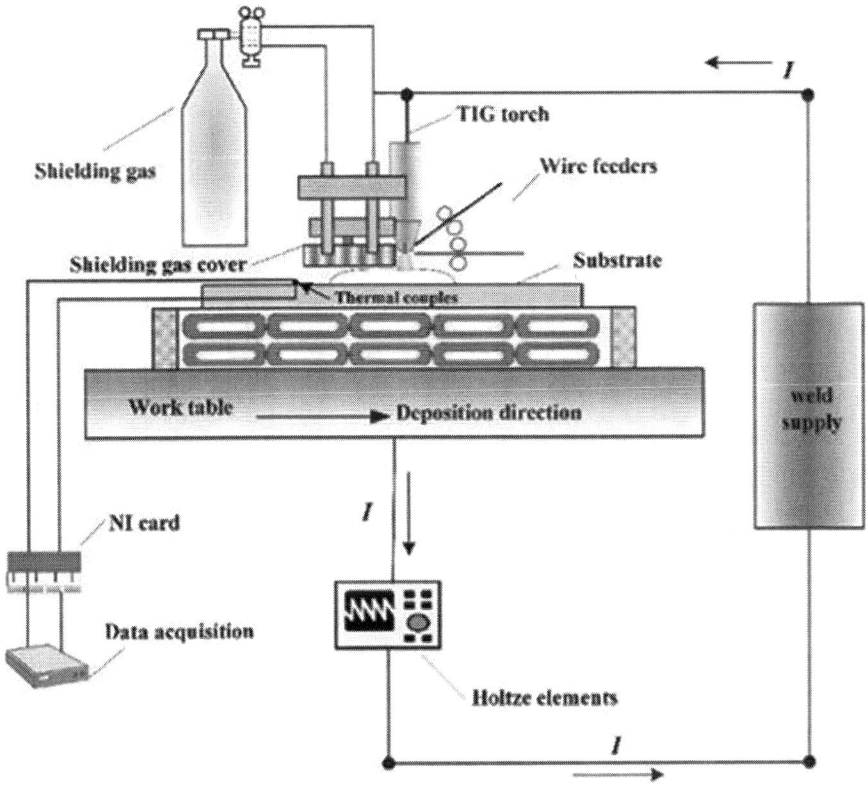

FIGURE 9.1 Wire additive manufacturing process: a process to produce metallic parts.

large structural components without being limited by the deposition rate. Also, a low cost of materials and equipment that does not compromise structural integrity is more associated with WAAM than WLAM [3]. For the WAAM technique, different materials, including titanium, steel, nickel-based superalloys, and Al, have been extensively utilised to produce AM products. These have been used extensively in biomedical implants, aerospace, automotive, and other industries [2–4]. Among all these materials, Al alloy has been the most commonly used to fabricate parts with the use of the WAAM technique, cutting across various lightweight applications [4]. The WAAM of Al alloys is widely used in aerospace and automotive industries for practical industrial applications [4]. The typical defects which often hamper the quality of the components fabricated by the WAAM include residual stresses, humping effect, oxidation, deformity, porosity, high surface waviness, cracking, and bad geometry accuracy in the final products [5, 6]. There is an urgent need for researchers to discover workable techniques that can permanently mitigate the effect of internal flaws formed at the initial stage of the manufacturing process in the WAAM system to correct these errors in the WAAM-processed parts. The structural integrity of parts produced by WAAM can be enhanced by certain manufacturing techniques. These include the optimisation of WAAM manufacturing parameters, post-processing treatments, and surface modification techniques, all of which have been researched extensively. In addition to the structural integrity, surface quality, mechanical properties, and microstructure properties of WAAM-processed Al alloys have been studied for the development of defect-free AM parts [7–10]. In recent times, a new technique has also been incorporated into the WAAM system: cold metal transfer (CMT). This technique utilises low heat input welding techniques to mitigate the effect of hot cracking and porosity defects in WAAM-processed Al parts [3]. By reducing heat input, the CMT process enhances weld quality by minimising distortion and spatter [5]. In this case, enhanced weld quality eliminates the need for post-processing techniques on the as-fabricated components.

Selecting suitable WAAM manufacturing parameters during the deposition process enhances the surface quality, mechanical properties, and microstructure properties of WAAM components [9]. The deposition manufacturing parameters set up for the WAAM system have to be fully optimised to achieve defect-free AM products. The full implementation of optimum parameters in the WAAM system will ensure that all targeted requirements of the final products during the printing process are extensively achieved. This chapter aims to critically assess the impact of WAAM manufacturing parameters and surface modification techniques on the WAAM-processed Al alloy parts. In this chapter, the significant benefit of WAAM manufacturing parameters and surface post-treatment used for additively manufactured metallic materials for the production of high-quality and defect-free AM products have been extensively reviewed and reported.

9.2 THE IMPACT OF WAAM MANUFACTURING PARAMETERS IN THE DEVELOPMENT OF DEFECT-FREE AM METALLIC PARTS

To produce a defect-free component for industrial settings, a set of manufacturing parameters must be thoroughly controlled and optimised. In the WAAM system, optimum process parameters of WAAM are required to ensure the final fabricated products meet the industrial component requirements. In WAAM processes, some key input parameters can be optimised to achieve optimal metallurgical characteristics in the fabricated components during the printing process. These parameters include welding strategy and travel speed [11, 12]; wire feed rate [11, 13]; working distance, arc current, and voltage [11, 14]; shielding gas flow rate [11, 14]; and deposition path strategy employed or path planning, part orientation, and inter-pass time, among others [15–20]. According to Klobcar et al. [21], linear heat input, the heat conductivity of feedstocks, and base plate are regarded as the main key input factors in process control. These factors ensure the process control of layer geometry, residual stresses, phase transitions, distortions, and prediction of cooling times. To regulate the effect of linear heat input in the arc procedure, travel speed, arc voltage, and current must be strictly observed during the printing process. The WAAM manufacturing parameters play a major role in the eventual outcome as regards the surface quality, microstructural properties, and mechanical properties of the part under consideration. In addition, the utilisation of optimum parameters during the printing process can mitigate the effect of defects in the fabricated components (see

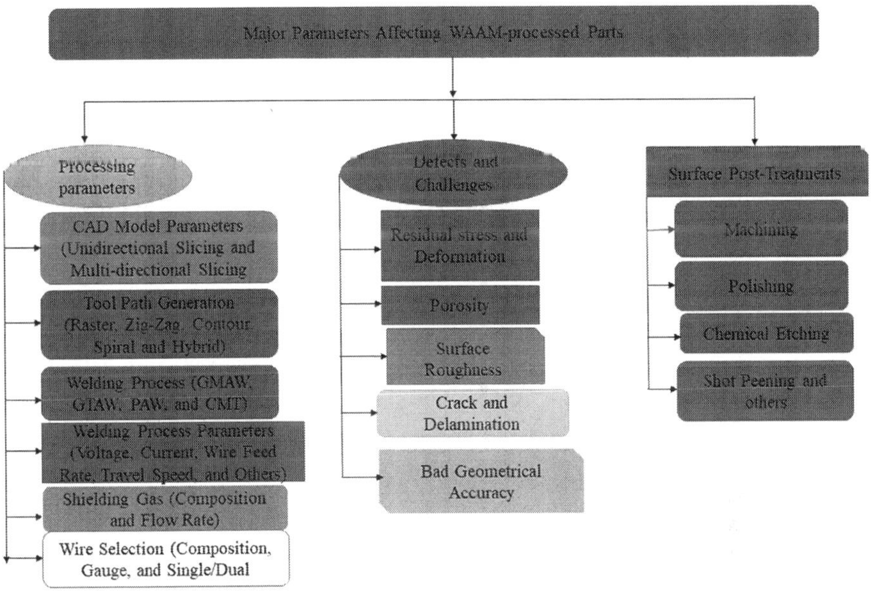

FIGURE 9.2 Wire additive manufacturing process: Major parameters affecting WAAM processing.

Figure 9.2). It has been established that wire feed speed and travel speed have a significant impact on surface waviness, as proven by Xiong et al. [19]; it is confirmed that layer height reduction through increasing travel speed or decreasing wire feed speed can minimise the surface waviness provided that the electric arc operates in a stable condition. Belhad et al. [20], via the design of experiments, studied the effect of the WAAM parameters on the surface quality as well as the part dimensional accuracy using the CMT-based WAAM process. This work affirmed that a decrease in the inter-pass time leads to a non-homogeneous collapse and the so-called mixed layer phenomenon.

The work by Dinovitzer et al. [17] discussed the effects of process parameters in tungsten inert gas (TIG)–based WAAM. In this work, the respective impact of key manufacturing parameters, such as rate of wire feed, travel speed, electric current, and shielding gas flow rate, were investigated. The responses of these parameters from an ANOVA test indicated that travel speed and current showed a major effect on the responses. It was further confirmed that a travel speed increase or a decrease in the current resulted in a reduction in melt-through depth and a corresponding increase in surface roughness. Concerning improving the geometric accuracy associated with these fabricated parts, Scetinec et al. [22] proposed an online layer control and in-process toolpath for gas metal arc welding (GMAW). This technique promotes improved geometric accuracy during the WAAM deposition of tall shell parts. In this work, deposition arc current and voltage were systematically changed to act as variables that controlled the layer height. It is affirmed that deposition arc current combined with re-slice control facilitates the manufacturing of parts with better final geometry fidelity compared to arc voltage. In related work, Chaurasia et al. [23]

discussed the significance of various process parameters, including travel speed, wire feed rate, heat input, deposition direction, and cold working. The work reported that travel speed has an inverse relationship with the humping defect, melt-through depth, and bead width. This means that controlling several processing manufacturing parameters can alleviate the effect of defects in the printed parts to achieve sound metallic components. Rauch et al. [24] considered parallel and oscillation deposition strategies for the fabrication of aluminium WAAM-fabricated components to enhance structural geometry, mechanical properties, and final products. Microstructure, micro-hardness, porosity, and mechanical properties were also examined for how they are affected by deposition strategy. However, it has been established that porosity and surface roughness remain the main issues affecting the full commercialisation of WAAM-processed aluminium alloys for industrial applications [24, 25]. Liang et al. [14] studied the behaviour of metal droplet transfer in WAAM of Al alloy. This work stated that the interaction of arc properties – namely, arc voltage and current – remains one of the influential factors that control characteristics of metal droplet transfer. Table 9.1 summarises previous works conducted on the aspect of improving the surface quality of WAAM-processed aluminium alloys.

9.3 THE INFLUENCE OF SURFACE TREATMENT TECHNIQUES ON THE IMPROVEMENT OF WAAM-PROCESSED AL ALLOYS

Al alloys produced by the WAAM process are usually subjected to post-process treatment to eradicate the internal flaws (porosity, distortions, cracks, residual stress, amongst others) arising from the printing of the parts and to improve the metallurgical properties [29, 32]. Surface treatment is widely regarded as being a necessary process to be carried out on the surface of as-fabricated materials to enhance surface properties and corrosion performance to improve its appearance. In AM techniques, the poor surface quality of WAAM Al parts has become one of the main challenges that have brought setbacks to the applicability of the WAAM products for industrial utilisation. A different category of surface treatment technique has been extensively studied (see Figure 9.3) and applied to AM metallic materials to proffer solutions to the internal flaws related to the surface roughness of as-fabricated materials to subdue these issues.

Omiyale et al. [33] described surface roughness as surface irregularities inherent in the fabricated objects that are directly associated with the manufacturing process. Olugbade [34] discussed the different surface treatment techniques, including ultrasonic shot peening, surface mechanical attrition treatment, machining, equal-channel angular processing, constrained grooved pressing high energy ball milling, among others, as a surface processing technique used for improving the corrosion resistance, surface properties alongside the mechanical properties of Al alloys. Li et al. [35] evaluated a hybrid WAAM milling process (HWMP) to improve the surface quality of stiffened panels. This work indicated that HWMP incorporated with optimal parameters enhances the surface properties of the material by 57% while improving the efficiency of the overall process by 32%, compared to conventional manufacturing processes. Table 9.2 presents previous works conducted on the aspect of surface treatment for improving the surface properties of Al alloys.

TABLE 9.1
The Impact of Manufacturing Parameters on the Improvement of Surface Quality of WAAM of Aluminum Alloys

Material	Manufacturing Parameter	(Mechanical) Properties Evaluated	Main Findings	Machine/Process	Ref.
Al-Mn-Fe-Si (ER5356)	i. Laser power: 0 W, 200 W, 400 W, 600 W ii. Deposition dimensions: 40 mm, 80 mm, 120 mm	Surface roughness	a. Increasing laser power up to a certain point improved surface roughness b. The ideal range for desirable surface roughness is between 200 and 400 W	Pulsed laser-assisted metal inert gas arc welding (L-M)	[25]
Al-Mg-Mn [Si-Fe-Zn-Ti-Cu-Cr] (AA5183)	i. Current: 230–240A ii. Arc Voltage: 32–33 V iii. Wire feed rate: 15 m/min iv. Wire nozzle distance: 16 mm v. Wire feed rate: 15 m/min vi. Travel speed: 12 mm/s vii. Shielding gas flow rate, nozzle diameter: 30 L/min (Argon), 18 mm	Presence of cracks Tensile strength Hardness	a. Multiple layer deposition led to cracking as subsequent passes affected reheated welds	Short pulse GMAW	[11]
Al-Cu-Mn-Mg-Zr-Ti-Zn (ER2319)	i. Travel speed: 150 mm/min, 250 mm/min, 350 mm/min, 450 mm/min ii. Welding current type: AC (50 Hz) iii. Wire feed speed: 2 m/min iv. Shielding gas flow rate: 18 L/min (Argon) v. Current intensity: 150A vi. Tungsten electrode diameter: 3.2 mm	Ultimate tensile stress Yield stress Elongation	a. Increasing laser power improved surface roughness b. Optimal roughness and forming quality were obtained at 350 and 450 mm/min	Gas tungsten arc welding (GTAW)	[26]

(Continued)

TABLE 9.1 (CONTINUED)

Material	Manufacturing Parameter	(Mechanical) Properties Evaluated	Main Findings	Machine/Process	Ref.
Al-Si-Fe-Cu-Mn (ER4043)	i. Wire properties based on Si content: wire 1 (5.22–5.38), wire 2 (4.98–5.53), wire 3 (5.34–5.8), wire 4 (5.87–6.14), wire 5 (5.55–6.83) ii. Tip working distance: 15 mm iii. Shielding gas flow rate: 25 L/min (Argon)	Surface quality Hardness distribution	a. Wire 1 and wire 2 led to cleaner, smoother welds b. Wire 3, 4, and 5 had more surface defects Scratches and pits were on wires 4 and 5, while wire 3 had processing/oxide residue	CMT – pulse advanced	[27]
Al-Cu-Mg-Mn-Ti-Zr-V-Zn-Si-Fe (AA2319)	i. Panel dimension: 200 mm × 120 mm ii. Wire source (manufacturer): denoted as A, B, and C iii. Gas flow rate: 20 L/min (Argon) iv. CMT mode: PADV, P v. Wire feed rate: 3.5 m/min, 4.5 m/min, 6 m/min vi. Travel speed: 0.3 m/min, 0.4 m/min, 0.6 m/min	Porosity Surface finish	a. Wire batch (as obtained from the manufacturer) has the most dominant effect on the porosity of welds, compared to CMT mode, wire feed speed, and travel speed b. The properties of the wire batch that affect porosity are hydrogen content, chemical composition, and wire surface finish. c. Auxiliary factors that could affect surface finish include the raw material source and the die through which wire drawing is done	CMT	[28]
Al-Cu-Mg-Zn-Mn-Si-Fe-Cr-Ti	i. Hot-wire current: 0A, 60A, 100A, 120A ii. Hot-wire voltage: 0 V, 2 V, 3 V, 3.5 V iii. Wire feed velocity: 100 cm/min, 180 cm/min, 300 cm/min, 350 cm/min iv. Scanning speed: 110 mm/min, 150 mm/min, 240 mm/min, 280 mm/min energy inpt::J/mm^3, 50.4J/mm^3, 43.1J/mm^3, 40.1J/mm^3	Ultimate Tensile stress Yield stress Elongation	a. Raising the hot-wire current led to a reduction in the total energy per unit volume and an increase in deposition rate b. Increased deposition rate led to a poor surface finish c. Using hot-wire reduced the porosity of the specimen	GTAW	[29]

Al-Mg-Mn-Cr-Cu-Fe-Si-Ti (ML5183)	i. Polarity variation: 10:10, 1:1 ii. Mean wire feed rate: 4.5 m/min, 5.2 m/min iii. Deposition rate: 0.81 Kg/h, 0.94 Kg/h iv. Current: 76/109A, 86/107A, 74/91A v. Mean voltage: 8.6 V, 10.6 V, 12.4 V vi. Travel speed: 12 mm/s vii. Energy input per unit length: 56 J/mm, 67 J/mm, 90 J/mm	Yield stress Tensile strength Fracture strain	a. Controlling heat input by varying polarity and pulse repetition sequence led to a significant improvement in the surface quality of the specimen b. Improved surface quality will likely positively impact the fatigue characteristics of the specimen surface as a result of reduced notch effects	CMT –advanced system	[12]
Al-Mg-Cr-Ti-Mn-x (ER5356)	i. Heat input: 6,160 J/mm, 7,248 J/mm, 8,568 J/mm, 10,197 J/mm, 12,250 J/mm ii. Current: 80A, 85A, 90A, 95A, 100A iii. Voltage: 13.2 V, 13.4 V, 13.6 V, 13.8 V, 14 V iv. Travel speed: 0.08 m/min, 0.09 m/min, 0.10 m/min, 0.11 m/min, 0.12 m/min v. Wire feed rate: 1.123 m/min, 1.263 m/min, 1.404 m/min, 1.544 m/min, 1.685 m/min	*Surface quality*	a. Heat input has a significantly high effect on the surface quality of specimens b. Extreme (too high and too low) temperatures negatively impact the surface quality of specimens	Square wave Argon arc	[30]
Al-Si-Mn-Mg-Cu-Fe-Zn-Ti (ER4043)	i. Boost current: 135A, 150A ii. Pulse duration: 2A, 5A, 9.5A iii. Wait current: 50A, 70A iv. Feed rate: 20 m/min, 25 m/min, 35 m/min, 37.5 m/min v. Short-circuit current: 40A vi. Linear current rise: 300A vii. Linear current decrease: 300A viii. Non-linear current rise rate: 0.1 ms	The geometric standard deviation for the width and thickness of deposited layers	The optimal range of the investigated parameters is presented below: a. Wait current (A): 40–70 b. Deposition rate (m/min): 20–45 c. Short-circuit current (A): 20–100 d. Boost current (A): 100–175 e. Pulse duration (ms): 1–2 f. Linear current rise (A): 300 g. Linear current decrease (A): 100–200	Cold metal transfer – advanced system (CMT-AS)	[13]

(Continued)

TABLE 9.1 (CONTINUED)

Material	Manufacturing Parameter	(Mechanical) Properties Evaluated	Main Findings	Machine/Process	Ref.
Al-Si-Fe-Cu-Mn-Mg-Zn-Ti (5356/5A06)	i. Deposition current: 135A ii. Deposition voltage: 18.5 V iii. Travel speed: 10 mm/s iv. Wire feed speed: 7.9 mm/s v. Shield gas flux: 20 L/min (Argon) vi. Distance from wire tip to contact tube: 15 mm vii. Angle between torch and substrate: 90^0 viii. Excitation current for magnetic field: 0A, 1.0A, 1.5A, 2.0A, 2.5A ix. Excitation frequency for magnetic field: 0 Hz, 30 Hz, 50 Hz, 70 Hz, 90 Hz, 110 Hz x. Average magnetic field intensity: 0 mT, 4.6 mT, 7.6 mT, 10.4 mT, 13.5 Mt	Tensile strength Yield strength Elongation	a. Raising the excitation current aids the dilation of the arc profile while increasing the excitation frequency does otherwise b. Given the appropriate setting of the parameters in (a), spray transfer and splatter are minimised, while arc stability is improved c. Increasing the excitation current of the mono-layer single-bead WAAM broadened bead width while also decreasing bead height and penetration depth. The opposite is the case when the excitation frequency is increased d. Electromagnetic stirring effect and better metal fluidity induced by the external magnetic field led to a reduction in the number and size of pores deposited	Cold metal transfer and electromagnetic field generation system (CMT+EMFGS)	[31]

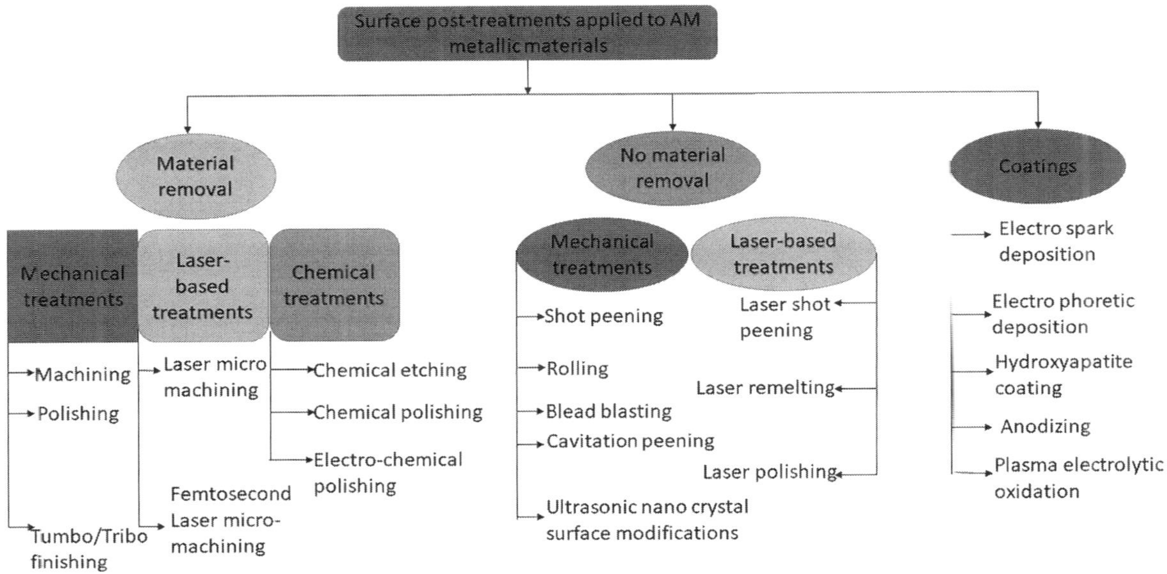

FIGURE 9.3 Classifications of surface post-treatment processing techniques applied to AM metallic materials.

TABLE 9.2
The Effects of the Surface Modification Techniques on the Surface Quality of AM Parts

Materials	WAAM Parameters	Surface Modification Process	Mechanical Properties Evaluated	Main Findings	Ref.
Al-Cu-Mn-Zr-Si-Mg-Zn-Ti (2319 Aluminum)	i. Welder type: TIG (tungsten electrode) ii. Wire diameter: 1.2 mm iii. Wire feed rate: 1.5 m/min iv. Shielding gas flow rate: 15 L/min (Argon) v. Travel speed: 0.3 m/min	Laser shock peening (LSP)	Tensile stress Yield strength Elongation	i. Microstructure refinement, leading to a grain reduction from 59.7μm to 46.7μm ii. Grain misorientation was minimised iii. Tensile and yield strength were enhanced by about 72% iv. No significant effect was found on ultimate tensile strength v. There was a drastic reduction in elongation vi. Surface-level micro-hardness was obtained	[36]
Al-Mg-Mn-Fe-Cu (ER5356)	i. Welder type: CMT ii. Current: 168A iii. Voltage: 16.9 V iv. Wire diameter: 1.2 mm v. Wire feed rate: 7.5 m/min vi. Travel speed: 0.6 m/min Electron beam densities: 5J/cm^2, 10J/cm^2, 15J/cm^2 vii. Accelerated voltage: 18 kV viii. Pulse duration: 200 µs	Electron beam surface treatment (EMST)	Tensile strength	i. EBST did not change the surface layer phase composition ii. At 10J/cm^2, the average tensile strength improved to 275.84 MPa iii. At an electron beam density of 15J/cm^2, the depth and thickness of the surface modification layer attained its maximum value iv. At an electron beam density of 10J/cm^2, the depth of these micro-cracks reached a minimum v. The loss of Mg and tensile strength at grain boundaries accelerates cracks beyond those boundaries	[37]

Al-Cu-Fe-Si-Mn-Mg-Zn-V-Ti-Zr-Be (Al 2319)	i. Welder type: CMT ii. Wire diameter: 1.2 mm iii. Wire feed rate: 6.5 m/min iv. Travel speed: 0.18 m/min v. Shielding gas flow rate: 25 L/min (Argon)	LSP	Hardness Yield strength	i. The yield strength and tensile strength of the specimen were increased by 151.2% and 13.7%, respectively ii. The method of LSP adopted influenced a surface layer of about 1.3 mm of the specimen iii. Pores were effectively collapsed due to stress concentration induced around them in the influenced layer iv. The fatigue life of the specimen was doubled due to the compressed residual stress, surface hardening, and defect elimination effects initiated by LSP	[38]
Al-Si-Fe-Cu-Mn-Mg-Zn-V-Ti-Zr (ER2319/Al 2219)	i. Welder type: CMT ii. Working current: 0A, 1A, 2.5A iii. Surface modification	Ultrasonic impact treatment (UIT)	Compressive residual stress Ultimate tensile strength Yield strength Elongation	i. UIT significantly reduced the number and size of pores in the specimen ii. It increased the dislocation density of the specimen iii. It has the potential to reduce the size of the precipitated phase iv. At UIT currents of 1A and 2.5A, the average compressive residual strength of the specimen's surface increased by 78.4% and 79.2%, respectively v. UIT currents of 1A and 2.5A increased ultimate tensile strength (10.8% and 11.9%) and yield strength (41.7% and 44.2%) but decreased elongation (35.7% and 44.8%)	[39]

(*Continued*)

TABLE 9.2 (CONTINUED)

Materials	WAAM Parameters	Surface Modification Process	Mechanical Properties Evaluated	Main Findings	Ref.
Al-Cu-Mg-Si-Mn-Zr-Fe-Ti (ER2319/Al 2024)	i. Welder type: TIG ii. Wire diameter: 1.2 mm iii. Wire feed rate: 2.4 m/min iv. Travel speed: 0.3 m/min v. Shielding gas flow rate: 18 L/min (Argon) vi. Arc mode current: 120A & 140A for variable polarity (VP) arc; 100A and 140A for ultrasonic frequency pulsed variable polarity (UFPVP) arc	Ultrasonic frequency pulsed arc (UFPA)	Hardness Tensile strength	i. UFPA refined the grain of the specimen However, it does not alter the random distributing properties of the grains ii. It increased the vertical tensile properties of heat-treated specimens while also improving its isotropy iii. UFPA reduced the porosity of the specimen iv. UFPA improved the micro-hardness and its uniformity in the specimens	[40]
Al-Cu-Mg-Mn-Si-Ti-Fe-Zn (ER2319/Al 2219)	i. Welder type: TIG; CMT ii. Wire diameter: 1.2 mm iii. Wire feed rate: 5.6 m/min iv. Travel speed: 5 mm/s v. Shielding gas flow rate: 25 L/min (Argon) vi. FSP rotation speed: 800 rpm vii. FSP travel speed: 200 mm/min	Interlayer friction stir processing (FSP)	Yield strength Ultimate tensile strength Elongation	i. WAAM + FSP initiated grain refinement ii. Tensile properties of the WAAM + FSP specimens were higher than those of ordinary WAAM iii. FSP got rid of pore defects in specimens and increased the fatigue limit by 43.3% iv. The tensile performance of the top-layer stir zone was better than that of the middle zone	[41]

Al-SI-Mg-Mn-Cu-Ti-Fe-Zn (ER4043/Al 6082-T6)	i. Welder type: TIG; CMT ii. Wire diameter: 1.2 mm iii. Wire feed rate: 5.6 m/min iv. Travel speed: 5 mm/s v. Shielding gas flow rate: 25 L/min (Argon) vi. FSP rotation speed: 800 rpm vii. FSP travel speed: 200 mm/min	Interlayer FSP	Ultimate tensile strength Yield strength Elongation	i. FSP reduced yield strength by 9.3%, improved the elongation by 108.7% and left the yield strength practically unchanged ii. The outcomes above were attributed to FSP deformation breaking up Si-rich eutectic phase grain refinement in the specimen iii. Fatigue limit was increased by 28% iv. The above behaviour, among others, may be attributed to the elimination of pores	[42]
Al-Si-Fe-Cu-Mn-Mg-Zn-Cr-Ti (AlSi5)	i. Welder type: CMT ii. Wire diameter: 1.2 mm iii. Wire feed rate: 3–5 m/min iv. Shielding gas flow rate: 4–10 L/min (Argon) v. Travel speed: 0.3 m/min	*Oxidation	-	i. Gas flow rate and nozzle-to-work distance have a high impact on the occurrence of oxidation anomalies ii. The heat input and wire feed rate have a medium impact on the occurrence of oxidation anomalies iii. The robot's travel speed has a low impact on the occurrence of oxidation anomalies	[32]

9.4 SUMMARY AND FUTURE OUTLOOK

In this chapter, the significant influence of WAAM manufacturing parameters and surface post-treatment techniques on the improvement of surface quality of WAAM-processed Al alloys have been extensively reviewed and documented. WAAM techniques are less desirable in certain applications due to the high heat input required, geometric inaccuracy of as-fabricated parts, porosity, and poor surface finish. This work has identified the significant influence of WAAM manufacturing parameters (including travel speed, wire feed rate, working distance, arc current, and voltage; shielding gas flow rate, nozzle diameter; and others) and surface post-treatment (including shot peening, chemical etching, machining, rolling, chemical polishing, among others) as a workable technique that can be used to eliminate the effect of defects formed during the manufacturing process. The following conclusions are drawn from the literature comprehensively reviewed to provide a background for future directions:

i. More research efforts in the aspect of hybrid manufacturing processes combining WAAM with the machining process are required.
ii. There is an urgent need for researchers to identify optimal parameters that can permanently mitigate the effect of internal flaws formed at the initial stage of the manufacturing process in the WAAM system to produce functional metallic parts.
iii. More research efforts are needed in the aspect of ultrasonic shot peening, surface mechanical attrition treatment, machining, equal-channel angular processing, constrained grooved pressing high energy ball milling, and other surface post-treatment methods to achieve the improved surface quality of metallic parts.
iv. Some research efforts have been made on wire feed rate, traverse speed, arc current and voltage [43]. More research efforts are still required in the following WAAM manufacturing parameters (nozzle diameter, shielding gas flow rate, working distance, part orientation, among others) and post-heat treatment parameters, including solution time, solution temperature, ageing temperature, ageing time, and others to achieve optimal parameters setting for the production of functional metallic parts.

REFERENCES

1. Srivastava M., Rathee S., Vivesk Patel V., Atul Kumar A., & Koppad P. G. (2022). A review of various materials for additive manufacturing: Recent trends and processing issues. *Journal of Materials Research and Technology*, 21, 2612–2641.
2. Barath Kumar M. D., & Manikandan M. (2022). Assessment of process, parameters, residual stress mitigation, post treatments and finite element analysis simulations of wire arc additive manufacturing technique, *Metals and Materials International*, 28, 54–111. https://doi.org/10.1007/s12540-021-01015-5
3. Zhang C., Li Y., Gao M., & Zeng X. (2018). Wire arc additive manufacturing of Al-6Mg alloy using variable polarity cold metal transfer arc as a power source. *Materials Science & Engineering, A*, 711, 415–423.

4. Omiyale B. O., & Farayibi P. K., (2020). Additive manufacturing in the oil and gas industries. *Analecta Technica Szegedinensia*, 14(1), 9–18. https://doi.org/10.14232/analecta.2020.1.9-18
5. Omiyale B. O., Olugbade T. O., Abioye T. E., & Farayibi P. K. (2022). Wire arc additive manufacturing of aluminium alloys for aerospace and automotive applications: A review. *Materials Science and Technology*, 38(7), 391–408.
6. Williams S. W., Martina F., Addison A. C., Ding J., Pardal G., & Colegrove P., (2015). Wire + arc additive manufacturing. *Mater. Sci. Technol.* 32(7), 641–647.
7. Bankong B. D., Abioye T. E., Olugbade T. O., & Zuhailawati H. (2022). Review of post-processing methods for high-quality wire arc additive manufacturing, *Materials Science and Technology*, 1–19. doi: 10.1080/02670836.2022.2110223
8. Casalino G., Karamimoghadam M., & Contuzzi N. (2023). Metal wire additive manufacturing: A comparison between arc laser and laser/arc heat sources. *Inventions*, 8, 52. https://doi.org/10.3390/inventions8020052
9. Jafari D., Vaneker T. H. J., & Gibson I. (2021). Wire and arc additive manufacturing: Opportunities and challenges to control the quality and accuracy of manufactured parts. *Materials & Design*, 202, 109471. doi:10.1016/j.matdes.2021.109471
10. Farayibi P. K., & Omiyale B. O. (2020). Mechanical behaviour of polylactic acid parts fabricated via material extrusion process: A taguchi-grey relational analysis approach. *International Journal of Engineering Research in Africa (JERA)*, 46, 32–44. https://doi.org/10.4028/www.scientific.net/jera.46.32
11. Horgar A., Fostervoll H., Nyhus B., Ren X., Eriksson M., & Akselsen O. M. (2018). Additive manufacturing using WAAM with AA5183 wire. *Journal of Materials Processing Technology*, 259, 68–74.
12. Klein T., & Schnall M. (2020). Control of macro-/microstructure and mechanical properties of a wire-arc additive manufactured aluminum alloy. *The International Journal of Advanced Manufacturing Technology*, 108, 235–244.
13. Gomez Ortega A., Corona Galvan L., Deschaux-Beaume F., Mezrag B., & Rouquette S. (2018). Effect of process parameters on the quality of aluminium alloy Al5Si deposits in wire and arc additive manufacturing using a cold metal transfer process. *Science and Technology of Welding and Joining*, 23(4), 316–332.
14. Liang Z., Li Jinglong L., Yi L., Jingtao H., Chengyang Z., Jie X., & Dong C. (2018). Characteristics of metal droplet transfer in wire-arc additive manufacturing of aluminum alloy. *The International Journal of Advanced Manufacturing Technology*, 99, 1521–1530.
15. Lockett H., Ding J., Williams S., & Martina F. (2017). Design for wire+ arc additive manufacture: Design rules and build orientation selection. *Journal of Engineering Design*, 28(7–9), 568–598.
16. Ayarkwa K. F., Pinter Z., Eimer E., Williams S., Ding J., & Suder W. (2021). Effect of the deposition strategy on al-cu alloy wire+ arc additive manufacture. *SVR – Materials Science and Engineering Technology*, 1(1), 28–34.
17. Dinovitzer M., Chen X., Laliberte J., Huang X., & Frei H. (2019). Effect of wire and arc additive manufacturing (WAAM) process parameters on bead geometry and microstructure. *Additive Manufacturing*, 26, 138–146.
18. Jin W., Zhang C., Jin S., Tian Y., Wellmann D., & Liu W. (2020). Wire arc additive manufacturing of stainless steels: A review. *Applied Sciences*, 10(5), 1563.
19. Xiong J., Li Y., Li R., & Yin Z. (2018). Influences of process parameters on surface roughness of multi-layer single-pass thin-walled parts in GMAW-based additive manufacturing. *Journal of Materials Processing Technology*, 252, 128–136.

20. Belhadj M., Kromer R., Werda S, & Darnis P. (2023). Effect of cold metal transfer-based wire arc additive manufacturing parameters on geometry and machining allowance. *The International Journal of Advanced Manufacturing Technology*, 24, 1–10.
21. Klobcara D., Balosb S., Busicc M., Duricd A., Lindica M., & Scetineca A. (2020). WAAM and other unconventional metal additive manufacturing technologies, *Advanced Technologies and Materials*, 45(2), 1–9.
22. Ščetinec A., Klobčar D., & Bračun D. (2021). In-process path replanning and online layer height control through deposition arc current for gas metal arc based additive manufacturing. *Journal of Manufacturing Processes*, 64, 1169–1179.
23. Chaurasia M., & Sinha M. K. (2021). Investigations on process parameters of wire arc additive manufacturing (WAAM): A review, *Mechanical Engineering*, 1–9. doi:10.1007/978-981-15-8542-5
24. Rauch M, Nwankpa U. V., & Hascoet J. (2021). Investigation of deposition strategy on wire and arc additive manufacturing of aluminium components, *Journal of Advanced Joining Processes*, 4, 100074.
25. Zhang Z., Sun C., Xu X., & Liu L. (2018). Surface quality and forming characteristics of thin-wall aluminium alloy parts manufactured by laser assisted MIG arc additive manufacturing. *International Journal of Lightweight Materials and Manufacture*, 1(2), 89–95.
26. Zhou Y., Lin X., Kang N., Huang W., Wang J., & Wang Z. (2020). Influence of travel speed on microstructure and mechanical properties of wire+ arc additively manufactured 2219 aluminum alloy. *Journal of Materials Science & Technology*, 37, 143–153.
27. Gu J. L., Ding J. L., Cong B. Q., Bai J., Gu H. M., Williams S. W., & Zhai Y. C. (2015). The influence of wire properties on the quality and performance of wire+ arc additive manufactured aluminium parts. *Advanced Materials Research*, 1081, 210–214.
28. Ryan E. M., Sabin T. J., Watts J. F., & Whiting M. J. (2018). The influence of build parameters and wire batch on porosity of wire and arc additive manufactured aluminium alloy 2319. *Journal of Materials Processing Technology*, 262, 577–584.
29. Fu R., Tang S., Lu J., Cui Y., Li Z., Zhang X. T., Chen Z., & Liu C. (2021). Hot-wire arc additive manufacturing of aluminum alloy with reduced porosity and high deposition rate. *Materials & Design*, 199, 109370.
30. Zeng J., Nie W., & Li X. (2021). The influence of heat input on the surface quality of wire and arc additive manufacturing. *Applied Sciences*, 11(21), 10201.
31. Zhao W., Wei Y., Zhang X., Chen J., & Ou W. (2022). Comparative investigation of wire arc additive manufacturing of Al-5% Mg alloy with and without external alternating magnetic field. *The International Journal of Advanced Manufacturing Technology*, 119(3–4), 1–17.
32. Hauser T., Reisch R. T., Breese P. P., Nalam Y., Joshi K. S., Bela K., Kamps T., Volpp J., & Kaplan A. F. (2021). Oxidation in wire arc additive manufacturing of aluminium alloys. *Additive Manufacturing*, 41, 101958.
33. Omiyale B. O., Ogedengbe I. I., Olugbade T. O., & Farayibi P. K. (2023). Corrosion performance of wire arc additive manufacturing of stainless steel: A brief critical assessment. *3D Printing and Additive Manufacturing*, 00, 1–14.
34. Olugbade T. O. (2023). Review: Corrosion resistance performance of severely plastic deformed aluminium based alloys via different processing routes. *Metal and Materials International*, 29(1), 1–29.
35. Li F., Chen S., Shi J., Tian H. & Zhao Y. (2017). Evaluation and optimization of a hybrid manufacturing process combining wire arc additive manufacturing with milling for the fabrication of stiffened panels. *Applied Sciences*, 7, 1–14.

36. Sun R., Li L., Zhu Y., Guo W., Peng P., Cong B., Sun J., Che Z., Li B., Guo C., & Liu L. (2018). Microstructure, residual stress and tensile properties control of wire-arc additive manufactured 2319 aluminum alloy with laser shock peening. *Journal of Alloys and Compounds*, 747, 255–265.
37. Geng Y., Panchenko I., Konovalov S., Chen X., & Ivanov Y. (2021). Effect of electron beam energy densities on the surface morphology and tensile property of additively manufactured Al-Mg alloy. *Nuclear Instruments and Methods in Physics Research Section B: Beam Interactions with Materials and Atoms*, 498, 15–22.
38. Jing Y., Fang X., Xi N., Chang T., Duan Y., & Huang K. (2023). Improved tensile strength and fatigue properties of wire-arc additively manufactured 2319 aluminum alloy by surface laser shock peening. *Materials Science and Engineering: A*, 864, 1–7.
39. Wang C., Li Y., Tian W., Hu J., Li B., Li P., & Liao W. (2022). Influence of ultrasonic impact treatment and working current on microstructure and mechanical properties of 2219 aluminium alloy wire arc additive manufacturing parts. *Journal of Materials Research and Technology*, 21, 781–797.
40. Cong B., Cai X., Qi Z., Qi B., Zhang Y., Zhang R., Guo W., Zhou Z., Yin Y., & Bu X. (2022). The effects of ultrasonic frequency pulsed arc on wire+ arc additively manufactured high strength aluminum alloys. *Additive Manufacturing*, 51, 102617.
41. Wei J., He C., Zhao Y., Qie M., Qin G., & Zuo L. (2023). Evolution of microstructure and properties in 2219 aluminum alloy produced by wire arc additive manufacturing assisted by interlayer friction stir processing. *Materials Science and Engineering: A*, 868, 144794.
42. He C., Wei J., Li Y., Zhang Z., Tian N., Qin G., & Zuo L. (2023). Improvement of microstructure and fatigue performance of wire-arc additive manufactured 4043 aluminum alloy assisted by interlayer friction stir processing. *Journal of Materials Science & Technology*, 133, 183–194.
43. Tawfik M. M., Nemat-Alla M. M., & Dewidar M. M. (2021). Enhancing the properties of aluminum alloys fabricated using wire arc additive manufacturing technique – A review. *Journal of Materials Research and Technology*, 13, 754–768.

10 Influence of Post-Processing Manufacturing Techniques on Wire Arc Additive Manufacturing of Ti-6Al-4V Components

A Brief Critical Assessment

Babatunde Olamide Omiyale
Federal University of Technology, Akure, Nigeria

10.1 INTRODUCTION

Additive manufacturing (AM) techniques have demonstrated the ability to solve the needs of the industry with the production of complex geometrical shapes as compared to subtractive manufacturing (see Figure 10.1). AM enables the production of complex geometrical shapes and mass-customised parts that are too hard or impossible to manufacture using subtractive manufacturing techniques, such as machining, casting, and forging (Omiyale & Farayibi, 2020). Wire arc additive manufacturing (WAAM) systems have been widely adopted for massive metallic parts for use in industrial applications (Kulikov et al., 2020). The benefit of WAAM over other production methods includes a high deposition rate, ease of manufacturing large parts, and low capital cost. The advantage of the two technologies (i.e., AM and conventional manufacturing techniques) remained fully untapped in the production of functional parts for industrial purposes. Combining different production chains under consideration of the whole AM processes will facilitate the reduction in the utilisation of post-processing. Titanium and its alloys have played a major role in biomedical, marine, oil and gas, transport, chemical, and aerospace industries due to their mechanical strength and ductility (Mashigo et al., 2021; Omiyale & Farayibi, 2020).

Due to increasing demands for the application of titanium and its alloys in the manufacturing industry, the utilisation of additive-manufactured titanium alloy parts has improved in industries. This has further called for the enhancement of AM products with the use of post-processing manufacturing techniques for the improvement of products fabricated by the WAAM process. In the same vein, the structural

 DOI: 10.1201/9781003363415-10

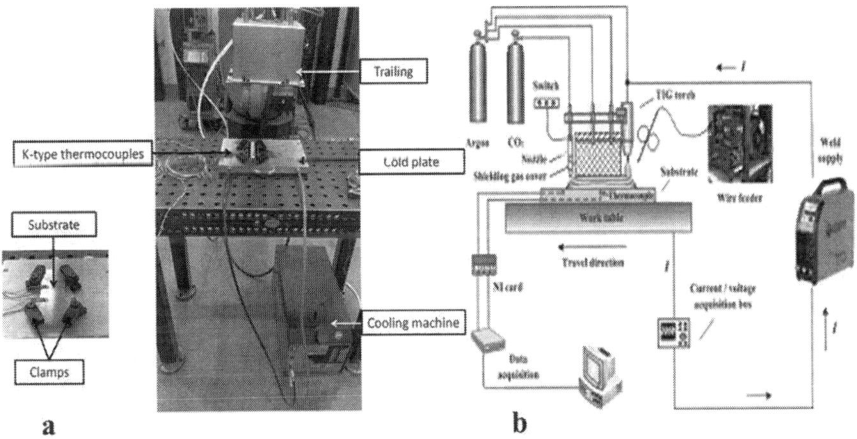

FIGURE 10.1 WAAM process. (Reprinted from Vazques et al., 2020; Wu et al., 2019.)

integrity of AM parts with the use of post-processing manufacturing techniques has led to an increase in the improvement of surface quality and mechanical and microstructure properties (Omiyale et al., 2022; Mashigo et al., 2021; Ding et al., 2015; Rosli et al., 2021). Among the AM technologies, the WAAM technique uses the wire feedstock in combination with a separate heat source to produce layers of metal on top of each other until a desirable metallic product is formed (Ding et al., 2015). The classification of WAAM-affiliated technologies can be done based on the source of energy that drives the process. These sources are primarily gas tungsten arc welding (GTAW), gas metal arc welding (GMAW), and plasma arc welding (PAW) (Rosli et al., 2021). Figures 10.2–10.4 present GTAW, GMAW, and PAW techniques used for the WAAM system.

Cold metal transfer (CMT), GTAW, GMAW, and PAW are the techniques used in the deposition of metallic layers in metal additive manufacturing (MAM) to create heat in WAAM.

Production of error-free components with the use of the WAAM system has become an arduous task due to the introduction of defects into the component during the printing process. This defect has brought great setbacks to the application of AM products in industries. In the AM technique, post-processing techniques have proffer solutions to mitigate the effect of common defects coming up during the printing of parts in the WAAM system. Some of these techniques (including the machining process, heat treatment techniques, hot isostatic pressing, and surface modification techniques, among others) have been effectively utilised to improve surface properties, mechanical properties, microstructure properties, and correct errors introduced during the fabrication of parts. This chapter systematically reviews the significance of post-processing methods adopted in the fabrication of defect-free components. It further considers the previous research efforts that have been made on the aspect of the machining process, heat treatment techniques, hot isostatic pressing, and surface modification techniques to improve the mechanical properties, microstructural characteristics, and measured quality of the surface of a WAAM-processed Ti-6Al-4V component.

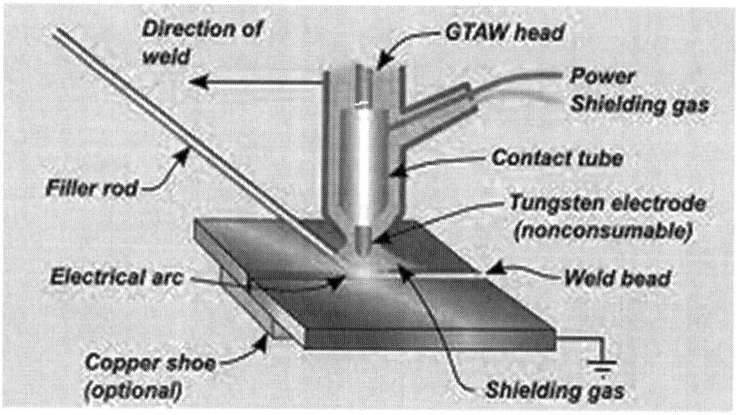

FIGURE 10.2 GTAW technique. (Reprinted from Kumar et al., 2017.)

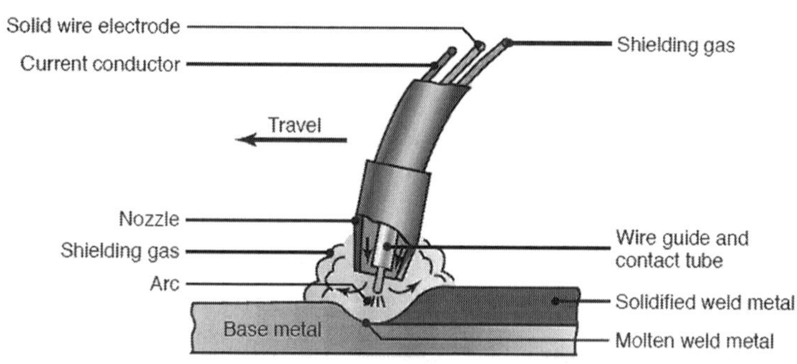

FIGURE 10.3 GMAW technique. (Reprinted from Bankong et al., 2022.)

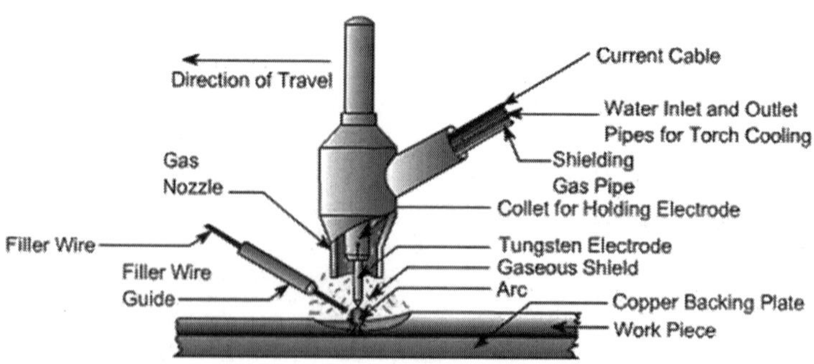

FIGURE 10.4 PAW techniques. (Reprinted from Bankong et al., 2022.)

10.2 PREVIOUS LITERATURE REVIEW RELATED TO WAAM OF TI-6AL-4V COMPONENTS

During the printing process, some internal defects in the printing products need to be mitigated, particularly for parts produced for extreme industrial purposes. The incur errors such as oxidation, porosity, lack-of-fusion imperfections, humping effect, distortion, cracking, and poor surface roughness defects occur in the WAAM system due to improper selection of manufacturing parameters, residual stress, and strain, uneven cooling rates of the melt pool, poor manufacturing set up approach, machine failures during the printing process, environmental factors, and others (Rosli et al., 2021; Caballero et al., 2019; Dávila et al., 2020; Zhang et al., 2016; Bankong et al., 2022). The defects produced in the WAAM system can be minimised by applying post-AM surface treatments and heat treatment techniques to tackle the challenges of shortcomings (such as cracking, porosity, oxidation, surface roughness, and others) associated with such components. Hence, more investigations are required to evaluate the influence of a different combination of post-processing manufacturing techniques to enhance the mechanical and microstructural properties of WAAM-processed Ti-6Al-4V components. Post-processing techniques contribute to the development of surface properties improvement to eradicate defect formation of AM components (Rauch & Hascoet, 2021; Zhang et al., 2016). Lesyk et al. (2021) applied laser shock peening (LSP), combined with multi-pin ultrasonic impact peening (UIP), water jet shot peening (WjSP), alongside water jet cavitation peening (WjCP) to improve surface quality, compressive residual stress values, hardness, and hardening depth in the composition surface layer. "Their results indicated that the surface micro-hardness improved by about 48%, 68%, and 80% after the combination of WjCP + LSP, WjSP + LSP, and UIP + LSP post-processing techniques in comparability with the as-built sample, respectively." A study by Rauch and Hascoet (2021) performed research on enhancing additive-manufactured surface properties with the use of post-processing techniques. This study discussed material removal or addition processes and surface properties modifications, including high-speed machining, waterjet machining, laser polishing, coating processes, and others, to improve material properties, corrosion, and wear resistance of Ti-6Al-4V parts produced by AM. The work by Martina et al. (2015) discussed the significance of inter-pass rolled in the enhancement of WAAM Ti-6Al-4V components. The authors concluded that inter-pass rolling of the as-fabricated components reduced their distortion and residual stress while also refining their microstructure, causing a transition from strongly columnar-to-equiaxed entities. In one study, Artaza et al. (2020) examined the impacts of heat treatment on components manufactured using the WAAM process. This work suggested that machining operation or etching as a post-processing technique can be well-utilised to further enhance the AM component surfaces. According to Williams et al. (2016), the authors informed that applying the in-process machining method to AM components to improve their surface properties is a better way of correcting errors or eliminating defects formation generated from the AM components. Baufeld et al. (2010) and Baufeld et al. (2011) worked on the microstructure characteristic of Ti-6Al-4V components. The influence of different kinds of heat treatment processes on microstructure and mechanical properties was reviewed. Based on the application of post-heat treatment techniques on the as-built samples, the microstructural properties of the Ti-6Al-4V components indicated fine α lamellae in the upper and coarse lamellae in the lowest region. Caballero et al. (2019) examined the influence of surface oxidation of titanium during WAAM deposition. This work affirmed that increasing the level of oxygen up to 4,000 ppm in the shielding environment significantly promotes an improved mechanical property and a decrease in elongation. In this section, an extensive literature review of post-processing techniques (i.e., heat treatment techniques, machining process, hot isostatic pressing, and surface modification techniques) which are used for enhancing the production of high-quality parts made of Ti-6Al-4V components have been fully documented. Post-processing manufacturing techniques have the potential to improve the surface quality and material properties of components produced using the WAAM process (Bankong et al., 2022; Ferro et al., 2020). It also helps to minimise and control the internal flaws formed during the printing of metallic parts. This section discusses four post-processing manufacturing techniques (including heat treatment techniques, machining process, hot isostatic pressing, and surface modification technologies) used for finishing parts made by the WAAM process.

10.2.1 Heat Treatment Techniques

This is one of the important techniques that have been well-utilised to alter and refine the material properties of WAAM-produced components (Vazquez et al., 2021; Mishurova et al., 2020). Essentially, heat treatment is carried out to enhance the microstructure properties, reform unwanted phases back in the matrix and relieve residual stresses, increase the mechanical strength, improve the hardness, and minimise the impact of cracking on additively manufactured parts (Ferro et al., 2020). Post-heat treatment techniques (which serve as a stress relief treatment) can be applied to the as-fabricated sample after printing to control the side-effect of residual stresses in AM-fabricated parts. This treatment is carried out at elevated temperature conditions to alleviate the risk of untimely failure and improve the surface quality and durability of these parts. Heat treatment techniques such as solid solution, quenching, and ageing treatment are usually applied to AM products that require robust mechanical strength (Jaber et al., 2022). Due to the low mechanical properties produced by AM technology (Ferro et al., 2020; Paghandeh et al., 2022), post-processing manufacturing techniques have played a key role in the refinement of AM components owing to low mechanical properties existing in as-built AM components (see Figure 10.5) (Ferro et al., 2020; Paghandeh et al., 2022; Bermingham et al., 2018; Lin et al., 2021; Hönnige et al., 2018). As reported by Qian et al. (2016), the heat treatment processes that are commonly used comprise sub-transus (i.e., α – β phase, just under ~ 996°C) and super-transus (i.e., β phase, just over ~ 996°C) regions for Ti-6Al-4V alloys modification. Figure 10.5 summarises the various techniques of post-processing utilised to refine parts fabricated by WAAM.

Artaza et al. (2020) investigated the impacts of the post-heat treatment on the enhancement of Ti-6Al-4V parts. After subjecting parts to annealing, microstructure analysis showed fine acicular α + β phases in the samples. Another study by Paghandeh et al. (2022) worked on the effect of phase improvement while conducting the thermomechanical processing of Ti-6Al-4V alloy. The authors concluded that

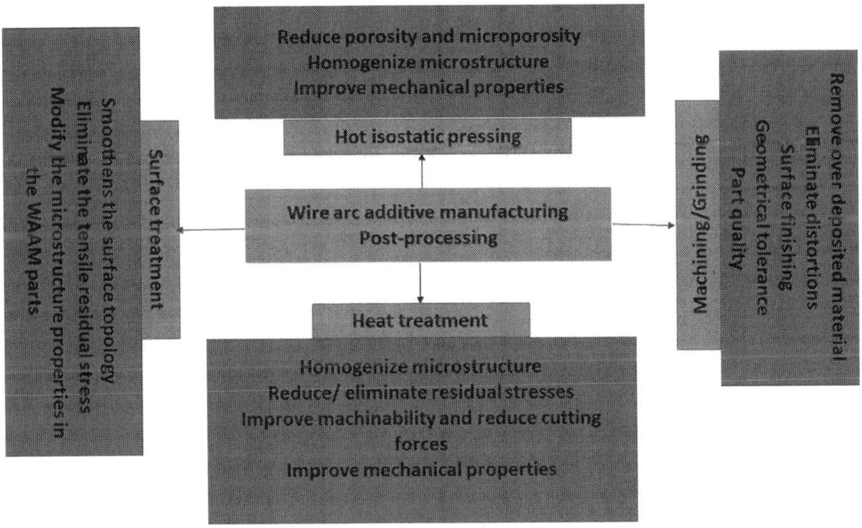

FIGURE 10.5 Post-processing techniques utilised to refine parts fabricated by WAAM.

ά martensite phase was achieved above β-transus temperature after water quenching by applying thermomechanical processing. Bermingham et al. (2018) optimised the post-heat parameters to achieve improved mechanical properties. The post-heat treatment process yielded a strength increase of up to about 12%. However, ductility was reduced by more than 30%. Lin et al. (2021) examined the outcome of post-heat treatment on the mechanical and microstructure properties of Ti-6Al-4V alloy manufactured by the WAAM. In this work, the yield strength and ultimate tensile strength of built parts were increased to about 12.8% and 3.33%, respectively, whereas tensile fracture elongation was reduced by 3.37%, compared to the as-built samples (Lin et al., 2021). Hönnige et al. (2018) also confirmed that the tensile residual stress found at the intersection of the built samples can be eradicated if as-built Ti-6Al-4V parts are left at an optimum temperature of 650°C for six hours (Hönnige et al., 2018). Going further, the study conducted by Vazquez et al. (2021) also examined the impact of thermal treatment on the enhancement of mechanical and microstructural properties of Ti-6Al-4V. After applying several thermal treatments at different temperatures, an appreciable lamellae ά phase was achieved in the processed materials. This led to an increase in the strength properties and ductility stability. According to Bermingham et al. (2018), thermal treatments have played a major impact in enhancing the growth of the phases and terminating martensitic structures to improve the stability of mechanical properties. Moreover, post-heat treatment is well-known as an important technique that not only enhances the mechanical properties of the final product by lamellar width coarsening and facilitating grain growth but also eradicates common defects brought about while the fabrication of the layers of manufactured parts is being done (Rubino et al., 2019). Furthermore, post-heat treatment may be effectively adopted in unlocking the optimal properties of materials, relieving residual stress and phases, and removing internal defects in the WAAM-processed

Ti-6Al-4V components (Rodrigues et al., 2019). One report says that both surface and heat treatment methods have been widely developed for AM technology to improve the mechanical properties of WAAM-manufactured components by refining the distribution of residual stress and material microstructure (Chi et al., 2020). Wang et al. (2020a) considered the effects of sub-transus heat treatment on the mechanical and microstructural properties of WAAM Ti-6Al-4V alloys. The heat treatment parameters of 930°C/1 h/WQ + 800°C/2 h/AC were adjudged to be optimal, according to their study. Elitzer et al. (2023) studied the impact of different process parameters and gas mixtures on the microstructure and mechanical properties of Ti-6Al-4V processed by WAAM. In this work, the significance of process parameters (including torch and wire feed speed) was examined. Wire feed speed was shown to have a significant effect on the geometry outline and mechanical properties of WAAM-processed Ti-6Al-4V alloys. In their subsequent application of heat treatment techniques, it was established that annealing after solution heat treatment inhibited mechanical properties, whereas basic stress relief promoted the improvement of mechanical properties. Dominguez et al. (2020) reported the growth of the α-case layer around the fabricated metallic component during the treatment of large and complex components derived from Ti-6Al-4V. This work acknowledged the need to eradicate the formation of the α-case layer on titanium alloy surfaces of this particular type of workpiece. For further research work, surface coating, chemical treatment, laser ablation, cathodic de-oxygenation, and other manufacturing techniques can also be explored to remove the effect of the α-case layer from titanium alloy surfaces. Tables 10.1 and 10.2 present the data on mechanical and microstructural properties of Ti-6Al-4V alloys manufactured using a WAAM system and improved with post-treatment techniques.

10.2.2 By Applying the Machining Process

WAAM has been described by Williams et al. (2016) as an advanced manufacturing technique that is being overwhelmed with the challenges of poor surface quality, which requires a finish machining pass to obtain a smooth surface. Due to imperfect shape and surface roughness inculcated in the built parts, post-processing through the machining process can be introduced in order to obtain a product with a smoother surface and greater dimensional accuracy. After the production of parts, the processing of parts with a machining process is required to remove unwanted deposited layers on the faces of built samples, leaving a desirable surface finish and eliminating distortions on the part of AM components. The work by Veiga et al. (2020) considered the manufacturing of Ti-6Al-4V by a combined PAW-WAAM and milling processes. Their study demonstrated that the up-milling machining process improved the quality of the final surface as compared to the down-milling machining process. In another study, Miyake et al. (2021) worked on the improvement of Ti-6Al-4V alloy using WAAM and high-temperature subtractive manufacturing techniques. Based on their report, the cutting force decreased by up to 20%, compared to carrying out the cutting at room temperature conditions, and they also obtained a good surface finish. Chernovol et al. (2021) researched the influence of WAAM parameters on the characteristics of as-deposited parts (i.e., initial surface waviness, flatness deviation,

TABLE 10.1
Mechanical Properties of Ti-6Al-4V Alloys Enhanced with Post-Treatment Techniques

Materials	Yield Strength (YS) (MPa)	Ultimate Tensile Strength (UTS) (MPa)	Elongation (EL) (%)	Post-Treatment Techniques	Ref.
Ti-6Al-4V	12.85% improvement in yield strength	Ultimate tensile strength of 3.33% was achieved.	Tensile fracture elongation was reduced by 3.37%.	Solid-solution treatment	Lin et al. (2021)
Ti-6Al-4V	Yield strength was slightly decreased by 8%	Ultimate tensile strength was slightly reduced by 8%.	Elongation increased by 70%.	Thermal treatment	Vazcuez et al. (2021)
Ti-6Al-4V		Increase ultimate tensile 886 ± 8 MPa, compared to 847 ± 12 MPa for the as-deposited specimen.	Increase elongation 16.6 ± 1.6%, compared to 12.2 ± 2.8% for the as-deposited specimen.	Solution and ageing method	Wang et al. (2020b)
Ti-6Al-4V		Ultimate tensile strength increases with an average of 996 MPa (in non-UPT state) to 1059 MPa (in UPT state).	Reduction in anisotropic percentages was achieved from 6 % (in the non-UPT state) to 0.8 % (in the UPT state).	Ultrasonic peening treatment (UPT)	Gou et al. (2020)
Ti-6Al-4V		It provides higher strength than these characteristics of WAAM Ti-6Al-4V alloys.	UNP-treated WAAM Ti-6Al-4V alloys demonstrate lesser elongation than those characteristics of WAAM Ti-6Al-4V alloys.	Ultrasonic Needle Peening Treatment (UNPT)	Yi et al. (2021)

TABLE 10.2
Microstructure Properties of Ti-6Al-4V Alloys Enhanced with Post-Treatment Techniques

Materials	Microstructure Properties	Post-Treatment Techniques	Machine	Ref.
Ti-6Al-4V	Microstructural tests showed satisfactory acicular $\alpha + \beta$ phases in the samples.	Heat treatment in a vacuum (argon-protected atmosphere) and air	WAAM	Artaza et al. (2020)
Ti-6Al-4V	Less significant influence on prior β grains was attained after applying post-heat treatment at the α+β phases.	Solid-solution treatment	WAAM	Lin et al. (2021)
Ti-6Al-4V	Fine dispersed granular αs with α martensite were attained after heat treatment.	Solution and ageing treatment	WAAM	Wang et al. (2020b)
Ti-6Al-4V	Solution treatment provided enhancement in grain size, which increases the strength of the built parts.	The solution, quenching, and ageing treatment	WAAM	Bermingham et al. (2018)
Ti-6Al-4V	Increasing heat treatment temperature generally increases the colony size and causes an improvement in average lamellar spacing (from about 75 nm to 220 nm).	Solution heat treatment	WAAM	Wang et al. (2019)
Ti-6Al-4V	Based on the effect of UPT on the processed materials, the recrystallised grains decreased while substructure and deformed grains increased.	Ultrasonic peening treatment	WAAM	Gou et al. (2020)
Ti-6Al-4V	In the UNP-treated specimens, fine α+α′ grains were observed with a low aspect ratio.	Ultrasonic needle peening (UNP) treatment	WAAM	Yi et al. (2021)

and wall thickness) and the machining process. The authors reported that deposition under shorter cooling times, such as inter-pass temperature and low heat input, led to better outcomes when the metrics for machining were assessed. From the studies so far, we deduce that there is a need for further critical evaluation of the effect of milling and welding parameters in the determination of the surface roughness of complex assemblies/components produced from AM-fabricated parts. In the case of welding, the parameters concerned include wire feed speed, travel speed, and inter-pass temperature. For milling, on the other hand, the parameters of interest include cutting velocity, feed rate, and axial depth of cut. In an investigation conducted by Fuchs et al. (2020), they modelled the essential machining allowance for post-processed WAAM-manufactured parts. Their study evaluated the alterations to surface roughness due to peripheral milling of such parts. The authors confirmed that taking off 12.5% of the highest surface could improve surface quality.

In the same vein, Li et al. (2017) used a regression model to estimate the effect of wire feed and travel speeds on the WAAM process. They also observed the effect of machining parameters on the resultant width and measured the roughness of the parts in question. It was shown that width and roughness demonstrated a high sensitivity to both the machining and WAAM process. Figure 10.6 presents the practical example of both deposited samples with the base plate without machining and after the machining operation.

Dvorak et al. (2021) combined the techniques of AM, metrology, and machining to manufacture AM objects to achieve the final quality part with surface finish and geometry accuracy that meet the industrial applications. Rauch and Hascoet (2022) also described the importance of high-speed machining (HSM) and waterjet machining as common finishing processes used for improving AM surfaces. The authors stated the objective of using HSM and waterjet machining as post-processing for WAAM is to eradicate the initial surface topology by removing unwanted layers covering the surface with another material. Farayibi et al. (2015) described waterjet machining as a post-processing technique suitable for producing a good surface finish on diverse engineering materials without experiencing any thermal damage. Bankong et al. 2022; Ding et al. 2019 described the friction stir process (FSP) as a new technique for increasing the surface quality of WAAM-processed components. The FSP is a technique that involves altering metal properties through high-intensity and focused plastic deformation. The deformation is effected using a non-consumable tool inserted in a workpiece and stirring the tool under pressure through the workpiece. Factors that affect the outcome of this method include the duration of the heating and cooling cycle, the extent of plastic deformation, and the loci of motion of the workpiece as it interacts with the tool. Despite the benefits of FSP, some limitations restrict the use of FSP as a post-processing manufacturing technique. In this case, the FSP is only useful for parts that are not complex in shape. In some AM components, geometric complexity parts do not allow a complementary machining process. For upgrading the FSP machine for better performance, the FSP efficiency

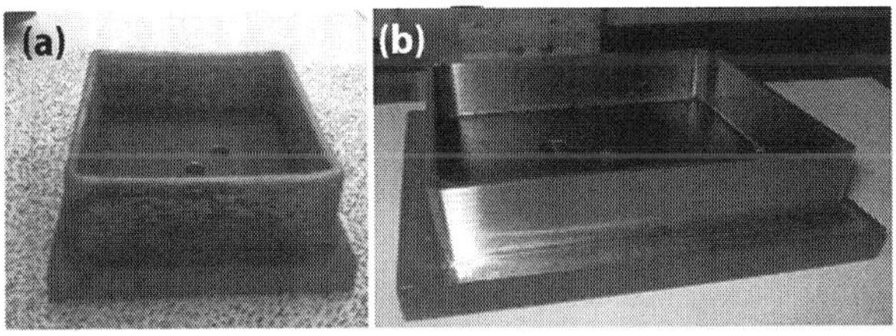

FIGURE 10.6 (a) Deposited sample with the base plate without machining. (b) Deposited sample with the base plate after machining. (Reprinted from Adebayo & Stephen, 2017.)

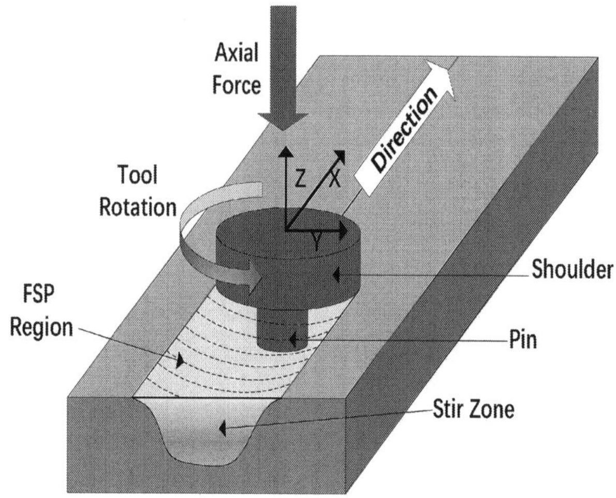

FIGURE 10.7 FSP.

can be significantly enhanced by integrating robotic arms capabilities into the tool-holder of the FSP machine for precision manoeuvring of the movement of the tool-holder and to allow the easy tool penetration for processing complex parts in an attempt to address this challenge. Another restriction of FSP application is that the microstructure of grains cannot be fully achieved or refined with a thicker section using FSP unless it is a thinner section of parts made by WAAM. A typical FSP is summarised in Figure 10.7.

10.2.3 Application of Hot Isostatic Pressing

The hot isostatic pressing (HIP) process utilises high gas pressure (usually with argon) to eliminate defects (such as pores, voids, and internal cracks) in AM-processed components (Bermingham et al., 2018). It has been established that fusion-related defects, such as porosity in additively manufactured products, can be controlled with the use of HIP techniques (Mclean et al., 2022). Benzing et al. (2019) proved that HIP treatment results in columnar-to-equiaxed phase transitions, as the morphology of β grains change to low aspect ratio α grains, thus removing microstructural heterogeneities. Moreover, Cai et al. (2016) studied how the densification behaviour, the microstructural phase evolution, the tensile strength, and the near-α high-temperature titanium allow manufacture through the HIP process. The authors demonstrated that the α_2 phase (Ti3Al) precipitates from the α matrix. Apart from the benefit of HIP technology in reducing porosity and healing microcracks, reports have it that the HIP process can reproduce defects such as distortion, potential surface contamination, and variability of distortion (Heaney & Binet, 2019). Authors reported that distortion can occur in the HIP chamber as a result of cooling gradients. Going by the report of the authors, this distortion can be fully mitigated by controlling the cooling rate of the work vessel. In terms of surface contamination, vapourising the contaminant from the surface using vacuum solution annealing has proven to be an effective mitigation strategy (Heaney & Binet, 2019). Hence, more advanced techniques can also be explored by the researchers to tackle these technical issues of distortion and surface contamination arising from the use of HIP technology. Another study by Mclean et al. (2022) explores the outcome of HIP and heat treatments on the porosity of WAAM Al 2319. This work established a 95% reduction in porosity after post-HIP heat treatment, lesser than in the as-fabricated samples; this result showed that HIP + heat treatments are a workable technique to minimise the effect of internal flaws in the WAAM process. Wang et al. (2022) studied the microstructure and tensile behaviour of TA15 alloy before different post-heating processes and after they had been conducted. It is affirmed in their study that HIP treatment exhibited a lamellar $\alpha+\beta$ microstructure with improved mechanical properties. Furthermore, Dolev et al. (2022) investigated the effect of the HIP technique on the as-fabricated samples. The HIPed samples showed an appreciable enhancement in the material's properties by increasing the mechanical properties of the manufactured parts. As stated by Dolev et al. (2022), HIP technology can be used to improve material properties by mitigating the level of porosity in AM products.

10.2.4 Application of Surface Modification Technologies

Titanium and its alloys have seen extensive adoption in aerospace and biomedical industries for the production of functional parts. In AM technology, the application of this material still requires surface modification technologies to make it fulfil its intended purposes. The metallic parts fabricated by the WAAM process can be significantly improved with the use of surface modification techniques. Surface modification techniques have been deployed in metals in order to enhance mechanical performance, surface integrity, and service life across the different fields of their application. Several post-AM surface treatments have been developed to enhance the quality of metallic parts, including LSP, shot peening (SP), UIP (John et al., 2021), laser polishing (LP), and others. Wang et al. (2020b) performed SP experiments on Ti-6Al-4V and reported its effect on crack propagation under fatigue loading conditions. In this work, the authors informed that SP inhibited the onset of fatigue-induced cracks. Mahmood et al. (2020) examined the impact of LSP on additively manufactured Ti-6Al-4V parts. The authors discovered that LSP was effective in controlling surface irregularities and local grain refinement. This also resulted in higher hardness values. The authors noted further that SP predisposes the piece to severe strain rates in thinner components. Rauch and Hascoet (2021) researched how to improve the surface properties of AM-processed pieces. Their work discussed the importance of applying the LP technique on AM components to improve and smoothen the surface topology. In another investigation, Avcu et al. (2020) studied a variety of properties associated with the surface and sub-surface regions of titanium alloy samples subject to different SP conditions (peening time and shot size). The authors demonstrated (as shown in Figure 10.8a) that the microstructure of unpeened samples showed a uniform distribution of $\alpha + \beta$ phases. As presented in Figure 10.8b–d, the authors found out that shot size is a major determinant of microstructural properties as well

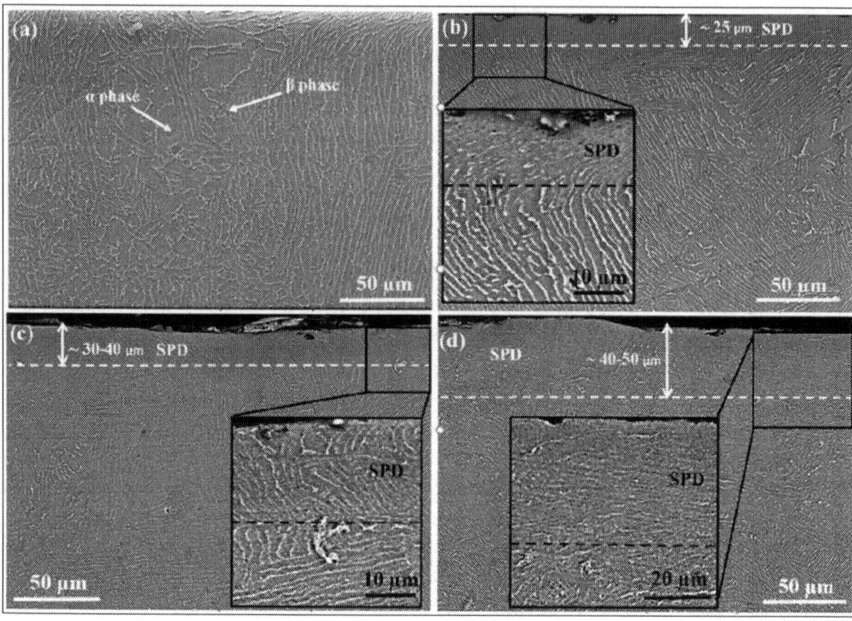

FIGURE 10.8 Microstructure investigation of Ti-6Al-4V: (a) reference sample; (b) S10, 5 min; (c) S60, 5 min; (d) S60, 15 min. (Reprinted from Avcu et al., 2020.)

as the hardness, while the peening time had a relatively less impact (Avcu et al., 2020). Wang et al. (2009) described SP as one of the useful surface treatment techniques that are widely applied in industry to enhance material properties and improve surface quality by inducing work hardening and compressive residual stress (CRS).

In terms of biomedical applications, metallic biomaterials are commonly used in the biomedical field (Wang et al., 2009). It serves as implant materials that have widely found their application in orthopaedic and dental (Xue et al., 2020). In one study, Xue et al. (2020) provided an overview of recent developments in SM methods of titanium alloys, specifically in biomedical applications. Some of the techniques discussed by the authors are plasma spray, sol-gel application, vapour deposition, and micro-arc oxidation, amongst others. It is stressed that bonding strength between the coating and metal substrate is significant for weight-bearing implants. Bond strength provides how powerfully each atom is strongly connected to another atom and how much energy is needed to disrupt the bond between the two atoms. To fully explore these methods, the authors suggested that the bonding strength of additively manufactured Ti-6Al-4V parts needs to be significantly improved. Based on the information gathered from the review, it is evident that surface modification technologies have a significant influence on the improvement of the surface topology of Ti-6Al-4V components for use in industrial applications (Rauch et al., 2021; Zhang et al., 2016; Wang et al., 2009; Xue et al., 2020; Ding et al., 2015). Hence, researchers still need to identify suitable surface treatment techniques and manufacturing process parameters that can bring out the best surface properties on AM-fabricated

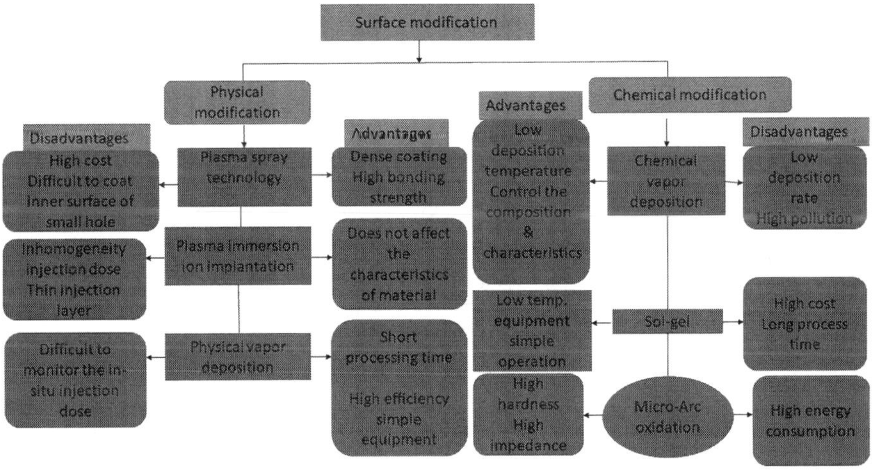

FIGURE 10.9 Surface modification techniques utilised to enhance surface properties of parts fabricated by WAAM.

parts. Figure 10.9 presents an overview of surface modification techniques with their advantages and disadvantages.

10.3 SUMMARY AND FUTURE OUTLOOK

In this chapter, it has been established that additively manufactured parts require post-processing implementation on the as-fabricated samples due to internal flaws generated during the printing of metallic parts (Bankong et al., 2022; Ferro et al., 2020). Post-processing is required to eliminate redundant surface material formed during the printing of Ti-6Al-4V components. The application of post-processing manufacturing techniques needs to be fully modified into the WAAM system to facilitate the production of defect-free components. In this comprehensive analysis of the relevant literature, AM post-processing manufacturing technique has been well-utilised to solve and address the problem of side effects (such as residual stresses, poor surface finish, oxidation, unfavourable mechanical properties, and others) generated from the WAAM process Omiyale et al. (2023). This review documented the effect of post-AM surface and heat treatment techniques on Ti-6Al-4V alloys processed using WAAM. Significant research and further understanding of aspects of post-heat treatment, machining process, HIP process, surface modification technologies, and other advanced mechanical surface treatment techniques towards improving surface characteristics of Ti-6Al-4V are required. For the fabrication of functional components, several aspects of this technology need to be thoroughly investigated and improved:

i. The influence of non-uniform properties created during the post-treatment process due to the presence of a non-uniform wall within the same built component needs to be fully optimised and controlled.

ii. More studies are needed to thoroughly assess the influence of the combination of different surface modification methods that have shown promise in the enhancement of mechanical and microstructural properties of additively manufactured Ti-6Al-4V to improve its operating life
iii. Optimum parameters (including wire feed speed, travel speed, inter-pass temperature, feed rate, axial depth of cut, and others) that could facilitate improvement in surface properties on AM-fabricated objects require comprehensive examination.
iv. More investigations are required from the researchers to identify the suitable surface treatment advanced techniques and process parameters that can bring out the best surface properties on AM Ti-6Al-4V fabricated parts.
v. More investigations are required to improve the bonding strength of additively manufactured Ti-6Al-4V parts significantly.
vi. More investigations are required to control the influence of the α-case layer generated on titanium alloy surfaces while handling large, complex geometries constituted of Ti-6Al-4V.

REFERENCES

Adebayo A. and Stephen J.T. (2017), "Effects of distortion on wire and arc additive manufacturing straight wall", *Journal of Building Construction, Planning, and Materials Research*, Vol. 2, No. 1, pp. 1–7.

Artaza A.T., Suárez A., Veiga F., Braceras I., Tabernero I., Larrañaga O. and Lamikiz A. (2020), "Wire arc additive manufacturing Ti6Al4V aeronautical parts using plasma arc welding: Analysis of heat-treatment processes in different atmospheres", *Journal of Materials Research and Technology*, Vol. 9, No. 6, pp. 15454–15466.

Avcu Y., Yetik O., Guney M., Iakovakis E., Sınmazçelik T. and Avcu E. (2020), "Surface, subsurface, and tribological properties of Ti6Al4V alloy shot peened under different parameters", *Materials*, Vol. 13, No. 19, p. 4363.

Bankong B.D., Abioye T. E., Olugbade T. O., Zuhailawati H., Gbadeyan O. O., et al. (2022), "Review of post-processing methods for high-quality wire arc additive manufacturing", *Materials Science and Technology*, pp. 1–19.

Baufeld B., Biest O.V.D. and Gault R. (2010), "Additive manufacturing of Ti–6Al–4V components by shaped metal deposition: Microstructure and mechanical properties", *Materials and Design*, Vol. 31, pp. 106–111.

Baufeld B., Brandl E. and Biest O.V.D. (2011), "Wire-based additive layer manufacturing: Comparison of microstructure and mechanical properties of Ti–6Al–4V components fabricated by laser-beam deposition and shaped metal deposition", *Journal of Materials Processing Technology*, Vol. 211, pp. 1146–1158.

Benzing J., Hrabe N., Quinn T., White R., Rentz R. and Ahlfors M. (2019), "Hot isostatic pressing (HIP) to achieve isotropic microstructure and retain as-built strength in an additive manufacturing titanium alloy (Ti-6Al-4V)", *Materials Letters*, Vol. 257, p. 126690.

Bermingham M. J., Nicastro L., Kent D., Chen Y. and Dargusch M. S. (2018), "Optimising the mechanical properties of Ti-6Al-4V components produced by wire + arc additive manufacturing with post-process heat treatments", *Journal of Alloys and Compounds*, Vol. 753, pp. 247–255.

Caballero A., Ding J., Bandari Y. and Williams S. (2019), "Oxidation of Ti-6Al-4V during wire + arc additive manufacture (WAAM)," *3D Printing and Additive Manufacturing*, Vol. 6, No. 1, pp. 91–98.

Cai C., Song B., Xue P., Wei Q., Yan C. and Shi Y. A. (2016), "Novel near α-Ti alloy prepared by hot isostatic pressing: Microstructure evolution mechanism and high-temperature tensile properties", *Materials & Design*, Vol. 106, pp. 371–379.

Chernovol N., Sharma A., Tjahjowidodo T., Lauwers B. and Van R. P. (2021), "Machinability of wire and arc additive manufactured components", *CIRP Journal of Manufacturing Science and Technology*, Vol. 35, pp. 379–389.

Chi J., Cai Z., Wan Z., Zhang H., Chen Z., Li L. and Guo W. (2020), "Effects of heat treatment combined with laser shock peening on wire and arc additive manufactured Ti17 titanium alloy: Microstructures, residual stress, and mechanical properties", *Surface and Coatings Technology*, Vol. 396, p. 125908.

Dávila J. L., Neto P. I., Noritomi P. Y., Coelho R. T. and da Silva J. V. L. (2020), "Hybrid manufacturing: A review of the synergy between directed energy deposition and subtractive processes", *The International Journal of Advanced Manufacturing Technology*, Vol. 110, Nos. (11–12), pp. 3377–3390.

Ding D., Pan Z., Cuiuri D. and Li H. (2015), "Wire-feed additive manufacturing of metal components: Technologies, developments, and future interests", *International Journal of Advanced Manufacturing Technology*, Vol. 81, pp. 465–481.

Ding Z., Fan Q. and Wang L. (2019), "A review on friction stir processing of titanium alloy: Characterization, method, microstructure, properties", *Metallurgical and Materials Transactions B: Process Metallurgy and Materials Processing Science*, Vol. 50, No. 5, pp. 2134–2162.

Dolev O., Ron T., Aghion E. and Shirizly A. (2022), "Effect of HIP defects on the mechanical properties of additive manufactured Ti6Al4V alloy", *Metals*, Vol. 12, pp. 1–11.

Dominguez L.A., Xu F., Shokrani A., Flynn J. M., Dhokia V. and Newman S. T. (2020), "Guidelines when considering pre & post-processing of large metal additive manufactured parts", *Procedia Manufacturing*, Vol. 51, pp. 684–691.

Dvorak J., Cornelius A., Corson G., Zameroski R., Jacobs L., Penney J. and Schmitz T. (2021), "Machining considerations for WAAM preforms", *ASPE 36th Annual Meeting*, Minneapolis, MN.

Elitzer D., Jäger S., Höll C., Baier D., Varga R., Zaeh M.F., Göken M. and Höppel H. (2023), "Development of microstructure and mechanical properties of TiAl6V4 processed by wire and arc additive manufacturing", *Advanced Engineering Materials*, Vol. 25, pp. 1–13.

Farayibi P. K., Abioye T. E., Murray J. W., Kinnell P. K. and Clare A. T. (2015), "Surface improvement of laser clad Ti–6Al–4V using plain waterjet and pulsed electron beam irradiation", *Journal of Materials Processing Technology*, Vol. 218, pp. 1–11.

Ferro P., Berto F., Bonollo F. and Romanin L. (2020), "Post welding heat treatment improving mechanical properties on Ti-6Al-4V", *Procedia Structural Integrity*, Vol. 26, pp. 11–19.

Fuchs C., Baier D., Semm T. and Zaeh M. F. (2020), "Determining the machining allowance for WAAM parts", *Production Engineering*, Vol. 14, pp. 1–9.

Gou J., Wang Z., Hu S., Shen J., Tian Y., Zhao G. and Chen Y. (2020), "Effects of ultrasonic peening treatment in three directions on grain refinement and anisotropy of cold metal transfer additive manufactured Ti-6Al-4V thin wall structure", *Journal of Manufacturing Processes*, Vol. 54, pp. 148–157.

Heaney D.F. and Binet C. (2019), "*Handbook of Metal Injection Molding*" Woodhead Publishing Series in Metals and Surface Engineering, United Kingdom, (2nd Edition), pp. 1–25.

Hönnige J., Colegrove P. A., Ahmad B., Fitzpatrick M., Ganguly S. and Lee T. (2018), "Design, residual stress, and texture control in Ti-6Al-4V wire+ arc additively manufactured intersections by stress relief and rolling", *Materials & Design*, Vol. 150, pp. 193–205.

Jaber H., Kónya J., Kulcsár K. and Kovács T. (2022), "Effects of annealing and solution treatments on the microstructure and mechanical properties of Ti6Al4V manufactured by selective laser melting", *Materials*, Vol. 15, No. 5, pp. 1–22.

John M., Kalvala P.R., Misra M. and Menezes P.L. (2021), "Peening techniques for surface modification: Processes, properties, and applications", *Materials*, Vol. 14, p. 384.

Kulikov A.A., Sidorova A.V. and Balanovskiy A.E. (2020), "Process design for the wire arc additive manufacturing of a compressor impeller", *IOP Conference Series: Materials Science and Engineering*, Vol. 969, p. 012098.

Kumar R., Mevada N. R., Rathore S., Agarwal N., et al. (2017), "Experimental investigation and optimization of TIG welding parameters on aluminum 6061 alloy using firefly algorithm", *IOP Conference Series: Materials Science and Engineering*, Vol. 225, No. 1, p. 012153.

Lesyk D.A., Soyama H., Mordyuk B.N., Stamann O. and Dzhemelinskyi V.V. (2021), "Combination of laser shock peening with cavitation, shot, and ultrasonic impact hardening for stainless steels surface characteristics improving", *Metallofizika i Noveishie Tekhnologii*, Vol. 44, No. 1, pp. 79–95.

Li F., Chen S., Shi J., Tian H. and Zhao Y. (2017), "Evaluation and optimization of a hybrid manufacturing process combining wire arc additive manufacturing with milling for the fabrication of stiffened panels", *Applied Sciences*, Vol. 7, No. 12, p. 1233.

Lin Z., Ya W., Song K. and Yu X. (2021), "Effect of post-heat treatment on step-by-step microstructure evolution and mechanical properties of Ti-6Al-4V alloy manufactured by wire and arc additive manufacturing", SSRN, p. 1–27.

Mahmood M.A., Chioibasu D., Rehman A., Mihai S. and Popescu A.C. (2020), "Post-processing techniques to enhance the quality of metallic parts produced by additive manufacturing", *Metals*, Vol. 12, No. 1, p. 77.

Martina F., Colegrove P. A., Williams S. W. and Meyer J. (2015), "Microstructure of interpass rolled wire + arc additive manufacturing Ti-6Al-4V component", *Metallurgical and Materials Transactions A*, Vol. 46, No. 12, pp. 6103–6118.

Mashigo L., Möller H. and Gassmann C. (2021), "Comparison of the mechanical properties of Grade 5 and Grade 23 Ti6Al4V for wire-arc additive manufacturing", *Journal of the Southern African Insitute of Mining and Metallurgy*, Vol. 121, No. 7, pp. 325–330.

Mclean N., Bermingham M.J., Colegrove P., Sales A., Soro N., Ng C.H., Dargusch M.S. (2022), "Effect of hot isostatic pressing and heat treatments on porosity of wire arc additive manufactured Al 2319", *Journal of Materials Processing Technology*, Vol. 310, p. 117769.

Mishurova T., Sydow B. and Thiede T. (2020), "Residual stress and microstructure of a Ti-6Al-4V wire arc additive manufacturing hybrid demonstrator", *Metals*, Vol. 10, pp. 1–15.

Miyake R., Sasahara H., Suzuki A. and Ouchi S. (2021), "Wire arc additive and high-temperature subtractive manufacturing of Ti-6Al-4V", *Applied Sciences*, Vol. 11, p. 9521.

Omiyale B.O. and Farayibi P.K. (2020), "Additive manufacturing in the oil and gas industries", *Analecta Technica Szegedinensia*, Vol. 14, No. 1, pp. 9–18.

Omiyale B.O., Ogedengbe I.I., Olugbade T.O., et al. (2023), Corrosion performance of additive manufacturing of stainless steel: A brief critical assessment. *3D Printing and Additive Manufacturing*, pp. 1–14, https://doi.org/10.1089/3dp.2022.0253

Omiyale B.O., Olugbade T.O., Abioye T.E. and Farayibi P.K. (2022), Wire arc additive manufacturing of aluminium alloys for aerospace and automotive applications: A review. *Materials Science and Technology*, Vol. 38, No. 7, pp. 391–408.

Paghandeh M., Zarei-Hanzaki A., Abedi H.R., Vahidshad Y. and Minarik P. (2022), "The correlation of c-to-a axial ratio and slip activity of martensite including microstructures during thermomechanical processing of Ti-6Al-4V alloy", *Journal of Materials Research and Technology*, Vol. 18, pp. 577–583.

Qian M., Xu W., Brandt M. and Tang H. P. (2016), "Additive manufacturing and postprocessing of Ti-6Al-4V for superior mechanical properties", *MRS Bulletin*, Vol. 41, No.10, pp. 775–784.

Rauch M. and Hascoet J. (2021), "Improving additive manufactured surface properties with post-processing techniques", *Advancing Precision in Additive Manufacturing*, St. Gallen, Switzerland, pp. 1–4.

Rauch M and Hascoet J. (2022), "A comparison of post-processing techniques for Additive Manufacturing components", *Procedia CIRP*, Vol. 108, pp. 442–447.

Rodrigues T. A., Duarte V., Miranda R. M., Santos T. G. and Oliveira J. P. (2019), "Current Status and Perspectives on Wire and Arc Additive Manufacturing (WAAM)", *Materials*, Vol. 12, No. 7, p. 1121.

Rosli N.A., Alkahari M.R., Abdollah M.F.B., Maidin S., Ramli F.R. and Herawan S.G. (2021), "Review on effect of heat input for wire arc additive manufacturing process", *Journal of Materials Research and Technology*, Vol. 11, pp. 2127–2145.

Rubino F., Scherillo F., Franchitti S., Squillace A., Astarita A. and Carlone P. (2019), "Microstructure and surface analysis of friction stir processed Ti-6Al-4V plates manufactured by electron beam melting", *Journal of Manufacturing Processes*, Vol. 37, pp. 392–401.

Vazquez L., Rodriguez M.N., Rodriguez I., Alberdi E. and Alvarez P. (2020), "Influence of interpass cooling conditions on microstructure and tensile properties of Ti-6Al-4V parts manufactured by WAAM", *Welding in the World*, Vol. 64, pp. 1377–1388.

Vazquez L., Rodriguez M.N., Rodriguez I. and Alvarez P. (2021), "Influence of Post-Deposition Heat Treatments on the Microstructure and Tensile Properties of Ti-6Al-4V Parts Manufactured by CMT-WAAM", *Metals*, Vol. 11, p. 1161.

Veiga F., Gil Del Val A., Suárez A. and Alonso U. (2020), Analysis of the machining process of titanium Ti6Al-4V parts manufactured by wire arc additive manufacturing (WAAM), *Materials*, Vol. 13, No 3, pp. 1–15.

Wang C.S., Li C.L., Chen R., Qin H.Z., Ma L., Mei Q.S. and Zhang G.D. (2022), "Multistep low-to-high temperature heating as a suitable alternative to hot isostatic pressing for improving laser powder bed fusion fabricated Ti-6Al-2Zr-1Mo-1V microstructural and mechanical properties. *Materials Science and Engineering*, Vol. 841, p. 143022.

Wang J., Lin X., Wang M., Li J., Wang C. and Huang W. (2020a), "Effects of subtransus heat treatments on microstructure features and mechanical properties of wire and arc additive manufactured Ti–6Al–4V alloy", *Materials Science and Engineering*, Vol. 776, p. 139020.

Wang J., Pan Z., Wei L., He S., Cuiuri D. and Li H. (2019), "Introduction of ternary alloying element in wire arc additive manufacturing of titanium aluminide intermetallic" *Additive Manufacturing*, Vol. 27, pp. 236–245.

Wang L., Lu W., Qin J., Zhang F. and Zhang D. (2009), "Influence of cold deformation on martensite transformation and mechanical properties of Ti–Nb–Ta–Zr alloy", *Journal of Alloys and Compounds*, Vol. 469, pp. 512–518.

Wang Y., Zhang Y., Song G., Niu W., Xu Z. and Huang C. (2020b), "Effect of shot peening on fatigue crack propagation of Ti6Al4V", *Mater. Today Communm*, Vol. 25, p. 101430.

Williams S. W., Martina F., Addison A.C., Ding J., Pardal G. and Colegrove P. (2016), "Wire + arc additive manufacturing", *Materials Science and Technology*, Vol. 32, No. 7, pp. 641–647.

Wu B., Pan Z., Chen G., Ding D., Yuan L., Cuiuri D. and Li H. (2019), "Mitigation of thermal distortion in wire arc additively manufactured Ti6Al4V part using active interpass cooling", *Science and Technology of Welding and Joining*, Vol. 24, Issue 5, pp. 484–494.

Xue T., Attarilar S., Liu S., Liu J., Song, L. and Tang Y. (2020), "Surface modification techniques of titanium and its alloys to functionally optimize their biomedical properties: Thematic review", *Frontiers in Bioengineering and Biotechnology*, Vol. 8, pp. 1–19.

Yi H., Kim J., Kim Y. and Shin S. (2021), "Improving mechanical properties of wire arc additively manufactured Ti-6Al-4V alloy by ultrasonic needle peening treatment", *Korean Journal of Materials Research*, Vol. 31, No 5, pp. 1–10.

Zhang J., Wang X., Paddea S. and Zhang X. (2016), "Fatigue crack propagation behaviour in wire+arc additive manufactured Ti-6Al-4V: Effects of microstructure and residual stress", *Materials & Design*, Vol. 90, pp. 551–561.

11 Mechanical Properties of Multi-layer Wall Structure Fabricated through Arc-Based DED Process

A Case Study

Ashish Yadav and Manu Srivastava

Hybrid Additive Manufacturing Laboratory, PDPM Indian Institute of Information Technology, Design, and Manufacturing, Jabalpur, India

Prashant K. Jain

FFF Laboratory, PDPM Indian Institute of Information Technology, Design, and Manufacturing, Jabalpur, India

Sandeep Rathee

National Institute of Technology Srinagar, Srinagar, India

11.1 INTRODUCTION

Today, additive manufacturing (AM) is an emerging technology to fabricate components with complex geometry and high accuracy [1]. AM is a process of material addition via a layer-by-layer approach as compared to the conventional machining processes of material removal for converting it into the desired shape and size. The AM CAD model is converted into an STL file; this file goes to the printer, which will print the three-dimensional solid component using the layered principle. Researchers have explored different materials in AM, i.e., polymers, steels, aluminium, biomaterials, etc. However, there is a need to explore many more materials to reduce dependence on the powder metallurgy industry and other conventional material manufacturing techniques and processes for material requirements. Currently, researchers are developing functionally graded materials (FGM) as per the specific requirements with customised mechanical properties, microstructure, and applications [2–4]. When

DOI: 10.1201/9781003363415-11

metallic materials are processed via AM techniques, it gives rise to a new and emerging AM domain called metal additive manufacturing (MAM). Recently, MAM has been growing too fast because of its application in marine, biomedical, oil and gas, automobiles, etc. The printing in MAM can be done through various processes such as powder bed fusion (PBF), direct energy deposition (DED), binder jetting, sheet lamination, etc.

Wire arc additive manufacturing (WAAM) is a special class of DED processes and has the capability of producing large components with a high deposition rate [5, 6]. GMAW-WAAM, GTAW-WAAM, and PAW-WAAM are different WAAM strategies. WAAM utilises input material as a wire electrode and a heat source for the electric arc to melt the material. The low buy-to-fly ratio and set-up cost make WAAM more economical than other MAM processes. WAAM has many applications in industries such as shipbuilding, oil and gas, automobile and aerospace, etc.

Alfredo et al. [7] printed a bimetallic wall with overlapping and sandwich structures of mild steel and SS316L and examined the mechanical properties and microstructure. The microstructure results showed that the wall has no lack of fusion nor micro-cracks or pores. The tensile results of the overlapping wall are higher than those of a sandwich wall, and both walls had similar hardness. Tomer et al. [8] examined the fatigue behaviour of ER70S-6 manufactured via WAAM and found a fatigue strength of 140MPa with a definite endurance limit. Feilong et al. [9] applied high-intensity ultrasound on build direction and found reduced grain size. Also, columnar grains were converted into equiaxed grains in the study. UTS and YS were found to increase in the vertical direction of the WAAM build. Sharma et al. [10] deposited the low-carbon steel material in a trochoidal (TR) toolpath and found grain with degenerate pearlite microstructure. The mechanical properties were uniform in the fabricated path with 494MPa strength and an average hardness value of 174 HV, respectively.

In this present study, a multi-layer wall structure of ER70S-6 was deposited on a mild steel substrate, and the mechanical properties of the deposited structure were studied.

11.2 MATERIAL SELECTION AND EXPERIMENTAL DETAILS

WAAM process is utilised to deposit a multi-layered structure of ER70S-6 material wire of 1.2 mm diameter as an electrode using an in-house developed WAAM set-up as represented in Figure 11.1. ER70S-6 wire has excellent welding properties such as low rust, stable deposition, less spatter, high welding efficiency, etc. ER70S-6 material has various applications in sheet metal, shipbuilding, and oil and gas industries, making these wires perfect for WAAM deposition. Table 11.1 shows the chemical composition of the ER70S-6 wire.

A heating source of Phoenix 351 power supply was used for arc creation, and a developed three-axis CNC system was used to move the WAAM torch in the required direction. Low-carbon steel ER70S-6 wire was used as a filler material with a mild steel substrate of 300 × 120 × 12 mm. Four clamps were fixed at 10 mm from the corner of the plate, and 80% argon and 20% CO_2 gases with a flow rate of 15L/min are considered in this study. The process parameters used in this study were obtained

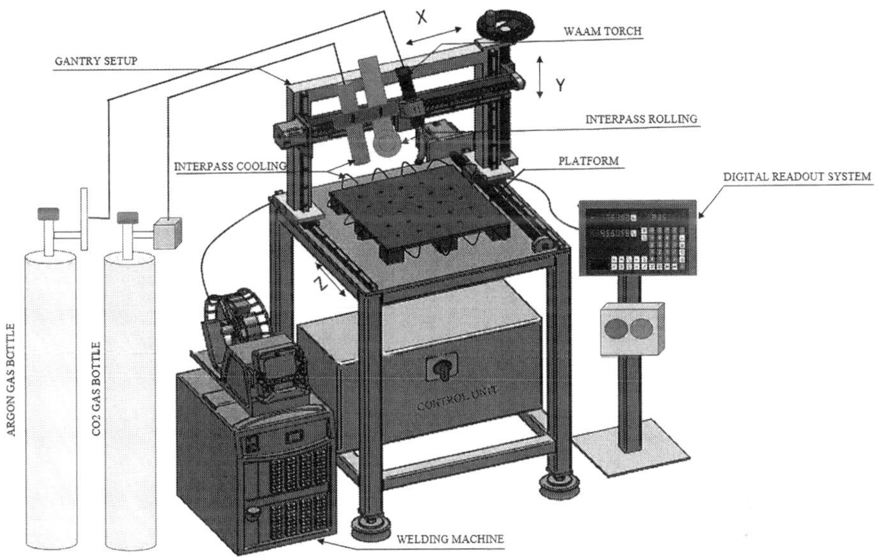

FIGURE 11.1 Schematic representation of 3D CAD model of the WAAM process.

TABLE 11.1
Chemical Composition of ER70S-6 Electrode and Substrate

	Chemical Composition (%)								
Elements (%)	**C**	**Mn**	**Si**	**Cr**	**Ni**	**Mo**	**S**	**P**	**Fe**
ER70S-6 (Electrode)	0.08	1.46	0.85	0.022	0.039	0.001	0.035	0.025	Balance
Substrate (Mild steel)	0.14	0.56	0.3	—	—	—	0.045	0.045	Balance

through various trial runs, as represented in Figure 11.2. Wire feed rate and voltage were kept at 4 m/min and 18V, respectively, during the complete deposition of the multi-layer wall structure. Other WAAM working process environments are given in Table 11.2. The inter-layer temperature of layers was maintained at 165°C, which was measured with the help of a Fluke IR thermometer.

Before deposition, the substrate plate was cleaned and ground, and then a multi-layer wall of 38 layers was deposited through the WAAM process at the obtained parameters. Samples were then prepared for the mechanical examination. Hardness and tensile tests were performed to see the mechanical behaviour of the deposited wall.

Dynamic servo control fully automatic machine was used to perform the hardness with 500g load applied for 10 sec. Hardness specimens were extracted from the multi-layered wall at three different locations, i.e., top zone, middle zone, and bottom zone, with the help of wire cut EDM machine. Tensile test specimens were prepared according to ASTM E8 standard. Similarly, tensile test specimens were extracted

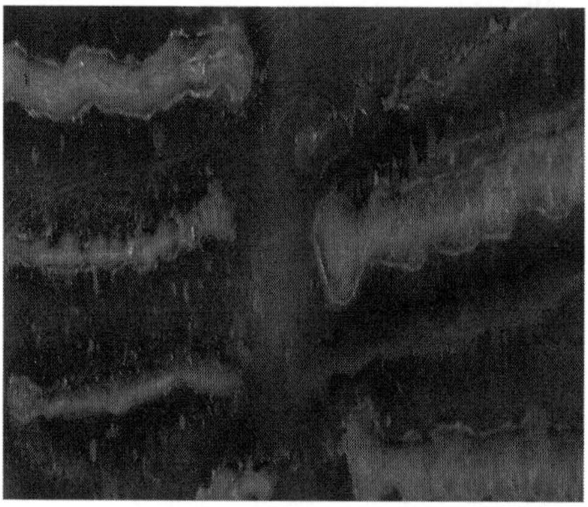

FIGURE 11.2 Trial runs to get process parameters.

TABLE 11.2
WAAM Process Environment

Deposition Mode	Direct Current Electron Positive (DCEP)
Gas composition	80% Argon + 20% CO_2
Electrode	ER70S-6
Standoff distance	10 mm
Electrode diameter	1.2 mm
WAAM torch speed	370 mm/min
Wire feed rate	4 m/min
Voltage	18 V

from the vertical and horizontal positions of the deposited multi-layer wall structure with the help of a wire EDM machine. A dynamic servo control fully automatic machine was used with a 0.25 mm/min rate for all tests.

11.3 RESULTS AND DISCUSSIONS

This study aims to experiment with the WAAM process to identify the most effective mechanical properties for WAAM-fabricated parts. Figure 11.3 shows the WAAM deposited wall of 38 layers.

The hardness specimen extracted from the deposited wall at the top, middle, and bottom zones, with the help of a wire cut EDM machine, is represented in Figure 11.4. Hardness tests were performed by dynamic servo control fully automatic machine with a load of 500g for 10 sec. Repeated heating or thermal cycle will cause the hardness variation in the deposited wall. The average microhardness values at the bottom, middle, and top regions are 174 HV, 168.36 HV, and 163.46 HV, respectively. Results show that the bottom zone has a maximum hardness value compared to the other two zones. The bottom-to-top zone hardness value shows a decreasing trend because of rapid solidification at the bottom zone, which decreases as layers increase. So that

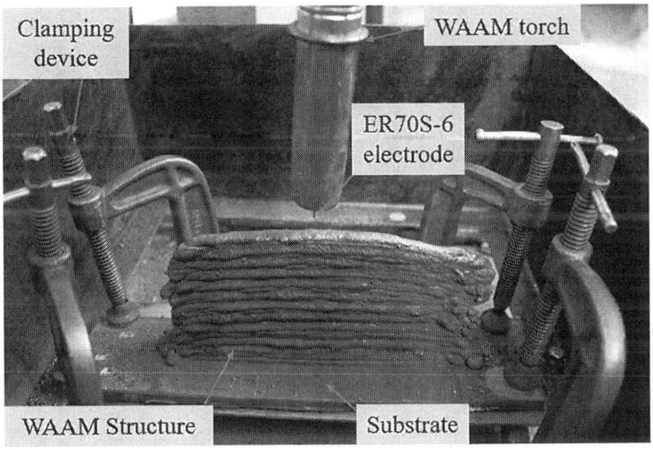

FIGURE 11.3 WAAM multi-layer wall fabricated structure.

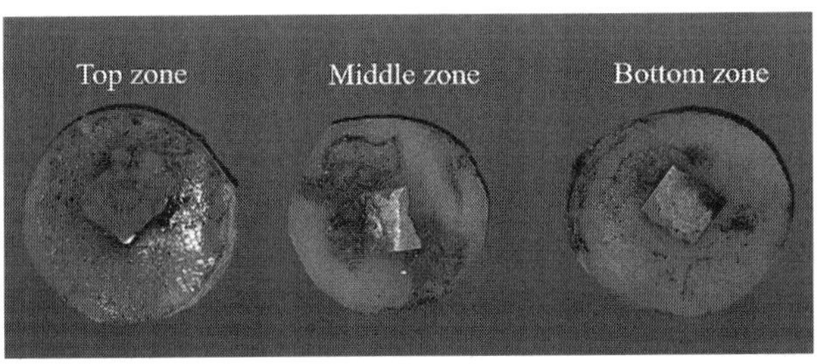

FIGURE 11.4 Hardness specimens.

fine grain will be developed. Residual stress and fine grains are responsible for hardness variation from the bottom to the top zone [11].

Similarly, samples were extracted from deposited WAAM structure through wire electrical discharge machining (wire EDM) at three horizontal and three vertical locations for the tensile test. Specimens were prepared as per ASTM E8 standard, as shown in Figure 11.5, and tests were performed on a dynamic servo control fully automatic machine. Ultimate tensile strength (UTS), yield strength (YS), and elongation percentage were considered for the tensile test. Before performing the tensile test on WAAM-fabricated specimens, wrought mild steel specimens were prepared, and the tensile test was performed on wrought mild steel, as represented in Figure 11.6. The average value obtained from the mild wrought steel is for the percentage elongation (El) of 28.1%, UTS of 437 MPa, and YS of 260.84 MPa, respectively. Results obtained from WAAM-fabricated multi-layered structures are represented in Table 11.3. The average tensile test values for elongation percentage, UTS, and YS are 31.96%, 514.94 MPa, and 330.64 MPa, respectively, as shown in Figure 11.7. The results obtained from the WAAM multi-wall structure are superior to wrought mild steel. The results of the current study are consistent with those of a previous study [12].

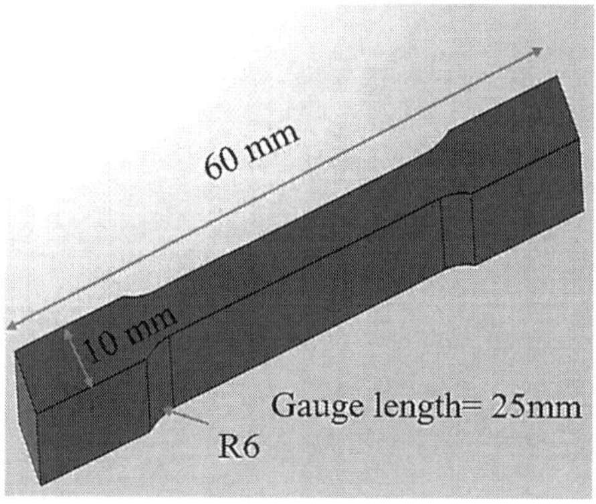

FIGURE 11.5 CAD model for dimension for the tensile specimen.

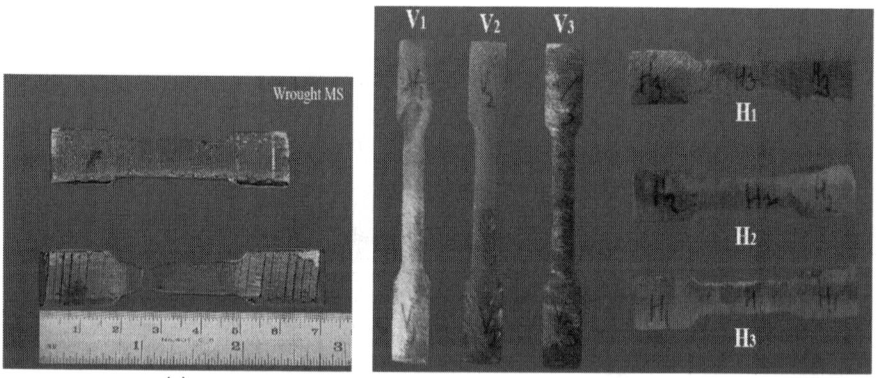

(a) (b)

FIGURE 11.6 Tensile specimens: (a) fractured tensile specimens of wrought mild steel and (b) WAAM-fabricated wall tensile specimens.

TABLE 11.3
WAAM-Fabricated Wall Structure Tensile Properties

Mechanical Properties	Elongation Percentage (EL %)			UTS (MPa)			YS(MPa)		
Location	Top	Middle	Bottom	Top	Middle	Bottom	Top	Middle	Bottom
WAAM ER 70S-6	36.23	32.46	27.2	524.04	516.45	504.34	336.67	329.36	325.91
Average value		31.96			514.94			330.64	
Wrought Mild steel		28.1			437			260.84	

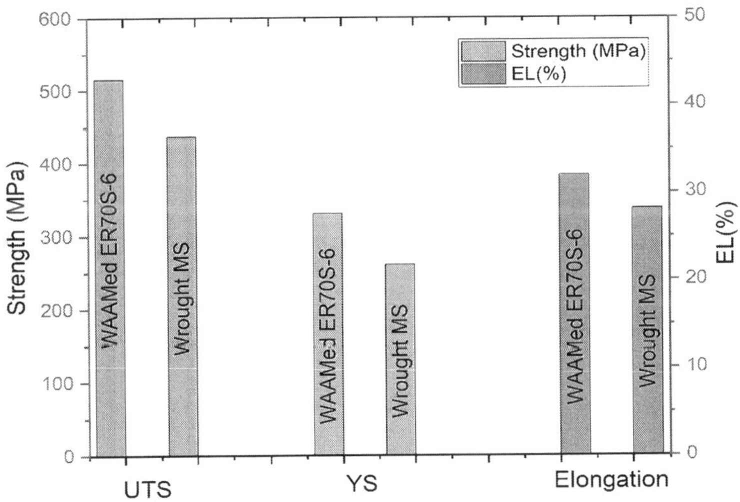

FIGURE 11.7 Tensile strength of WAAM-fabricated wall structure.

Tensile strength at the bottom zone has less value than other zones because the cooling rate at the bottom is highest; whenever layers increase, the heat source distance from the bottom zone will also be increased. At the top zone, the coarse microstructure will be there because of less cooling rate, and repetitive heating will also occur at the top zone. However, the tensile strength at all three zones has significantly fewer differences and shows a uniform behaviour of WAAM-fabricated structure.

Hall patch equation shows the effect of grain size on mechanical properties.

$$\sigma_y = \sigma_o + kd^{-1/2} \tag{11.1}$$

Where, σ_y is yield stress at a given direction, d is average grain size, and σ_o and k are the material constant.

11.4 CONCLUSION

In this study, a multi-layer wall was fabricated via WAAM for ER70S-6 with optimised WAAM process parameters. Hardness and tensile tests were performed to identify the most effective mechanical properties of the fabricated wall. The following are the outcomes of this research:

- The WAAM-fabricated wall shows the complete fusion between layers with no cracks and reduced porosity.
- The WAAM-fabricated multi-layered structures have average tensile test values for elongation percentage, UTS, and YS of 31.96%, 514.94 MPa, and 330.64 MPa, respectively.
- The average microhardness values at the bottom, middle, and top regions are 174 HV, 168.36 HV, and 163.46 HV, respectively.

- The present study's optimum process parameters of the WAAM process are torch speed of 370 mm/min mm/min, wire feed rate of 4 mm, and voltage of 18V, respectively.
- The mechanical properties of WAAM-fabricated multi-layer wall structures fall into the standard category of mild steel and will generally help produce components with superior mechanical properties.

This research might be expanded to examine the mechanical characteristics of additional materials and alloys, such as stainless steel, titanium, and nickel alloys, when manufactured using the DED technique. This might lead to a better knowledge of the DED process and assist in identifying the optimal materials and process variables for specific applications. The mechanical properties of the produced components can also be estimated using prediction models based on the process parameters and material characteristics. This may facilitate part design optimisation and speed up production while simultaneously lowering the cost of manufacturing.

NOMENCLATURE

ASTM	American Society for Testing and Materials
MAM	Metal Additive Manufacturing
WAAM	Wire Arc Additive Manufacturing
GMAW	Gas Metal Arc Welding
GTAW	Gas Tungsten Arc Welding
PAW	Plasma Arc Welding
UTS	Ultimate Tensile Strength
YS	Yield Strength
EL	Elongation

REFERENCES

1. Mu H, He F, Yuan L, et al. (2023). Toward a smart wire arc additive manufacturing system: A review on current developments and a framework of digital twin. *J Manuf Syst* 67: 174–189.
2. Badoniya P, Yadav A, Srivastava M, et al. (2022). *Fabrication of Functionally Graded Materials (FGMs) Via Additive Manufacturing Route BT – High-Performance Composite Structures: Additive Manufacturing and Processing*. Springer Singapore, Singapore, pp. 191–213.
3. Yadav A, Badoniya P, Srivastava M, et al. (2021). 10 Functionally Graded. *Funct Graded Mater Fabr Prop Appl Adv* 217–230.
4. Srivastava M, Rathee S, Patel V, et al. (2022). A review of various materials for additive manufacturing: Recent trends and processing issues. *J Mater Res Technol* 21: 2612–2641. https://doi.org/10.1016/j.jmrt.2022.10.015
5. Yadav A, Srivastava M, Jain PK, Rathee S, et al. (2023). Investigation of bead morphology and mechanical behaviour for metal inert gas welding-based WAAM in pulsed mode metal transfer on 316LSi stainless steel. *J Adhes Sci Technol*: 1–32. https://doi.org/10.1080/01694243.2023.2241642

6. Yadav A, Srivastava M, Jain PK, et al. (2023). Experimental investigation on mechanical behaviour of austenitic stainless steel fabricated through wire arc additive manufacturing. *Mater Today Proc* https://doi.org/10.1016/j.matpr.2023.07.210
7. Suárez A, Panfilo A, Aldalur E, et al. (2022). Microstructure and mechanical properties of mild steel-stainless steel bimetallic structures built using Wire Arc Additive Manufacturing. *CIRP J Manuf Sci Technol* 38: 769–773. https://doi.org/10.1016/j.cirpj.2022.06.018
8. Ron T, Levy GK, Dolev O, et al. (2020). The effect of microstructural imperfections on corrosion fatigue of additively manufactured ER70S-6 alloy produced by wire arc deposition. *Metals (Basel)* 10: 98.
9. Ji F, Hu Z, Qin X, et al. (2023) Improving microstructure and mechanical properties of thin-wall part fabricated by wire arc additive manufacturing assisted with high-intensity ultrasound. *J Mater Sci* 58: 2381–2395. https://doi.org/10.1007/s10853-023-08155-6
10. Sarma R, Singh AK, Kapil S, et al. (2023). Evolution of near homogenous mechanical and microstructural properties in wire-arc based directed energy deposition of low carbon steel following trochoidal trajectory toolpath. *J Mater Process Technol* 315: 117921.
11. Afrouzian A, Groden CJ, Field DP, et al. (2022). Additive manufacturing of Ti-Ni bimetallic structures. *Mater Des* 215: 110461.
12. Yadav A, Srivastava M, Jain PK, et al. (2024). Functionally graded deposition of dissimilar steel (316LSi and ER70S-6) fabricated through twin-wire arc additive manufacturing. *Mate Lett* 354: 135395 https://doi.org/10.1016/j.matlet.2023.135395.

Index

Pages in *italics* refer to figures and pages in **bold** refer to tables.

P

R

S

T

U

V

W

Y

9781032419145